RSPB BIRDING YEAR

Seasonal tips and activities to learn about bird behaviour

Siân Duncan and Dominic Couzens

BLOOMSBURY WILDLIFE

LONDON · OXFORD · NEW YORK · NEW DELHI · SYDNEY

BLOOMSBURY WILDLIFE
Bloomsbury Publishing Plc
50 Bedford Square, London, WC1B 3DP, UK
29 Earlsfort Terrace, Dublin 2, Ireland

BLOOMSBURY, BLOOMSBURY WILDLIFE and the Diana logo
are trademarks of Bloomsbury Publishing Plc

First published in the United Kingdom 2024
Text © Siân Duncan and Dominic Couzens 2024
Photos © 2024 as credited on page 220

A catalogue record for this book is available from the British Library.

Library of Congress Cataloguing-in-Publication data has been applied for.

ISBN: 978-1-3994-1342-8
ePub: 978-1-3994-1343-5
ePDF: 978-1-3994-1344-2

2 4 6 8 10 9 7 5 3 1

Designed by Austin Taylor

Printed and bound in China by C&C Offset Printing Co., Ltd.

To find out more about our authors and books visit www.bloomsbury.com
and sign up for our newsletters.

Published under licence from RSPB Sales Limited to raise awareness of the RSPB
(charity registration in England and Wales no 207076 and Scotland no SC037654).
For all licensed products sold by Bloomsbury Publishing Limited, Bloomsbury Publishing
Limited will donate a minimum of 2% from all sales to RSPB Sales Ltd, which gives all
of its distributable profits through Gift Aid to the RSPB.

Contents

JANUARY

1

Hungry birds

Winter is a great time of year to get into birdwatching.
A lack of natural resources means that should you put a little
food out, it won't be long before you get birds flocking.

IT'S A COLD, CRISP January morning. Dawn is slowly breaking, causing the snow to sparkle and shimmer as the sun spreads its weak warmth. It's beautifully quiet. A world muffled under a white blanket. Then, from high on a bare branch, a Robin sings and the sweet tumble of trills provides an instant hit of joy. While I know that the song was not really meant for me, it also acts as a reminder to replenish the feeders. With weather this cold, this little Robin and many other birds like him could do with a helping hand.

← A Bullfinch finds seeds in the snow.

↓ Bird tables can tempt in hungry Jays.

Winter can pose a particular problem for birds as their natural food sources become scarcer. Insects are hunkered down, hibernating and inactive, and the berries and seeds that were once so bountiful dwindle to remnants as the season progresses.

Focused on feeding

For a bird in winter, life is all about finding food for survival. Birds have a relatively high body temperature of around 40°C – higher than a human's 37°C – and a high metabolism. Just keeping warm requires a huge amount

→ Goldfinches squabble over nyjer seeds.

of energy. Regular food sources, such as insects and other invertebrates, as well as nuts, seeds and berries, are harder to find and the days are much shorter too, meaning birds have to squeeze their food-finding into fewer daylight hours.

All things considered, it's perhaps not so surprising that birds need to spend most of their waking hours hunting for food in winter. A Great Tit, for example, will spend 75 per cent of its time hunting for food, with the smaller Blue Tit spending 85 per cent and the tiny Coal Tit needing to devote 90 per cent of its waking hours to feeding.

Lending a helping hand

By now, you are hopefully resolved to help our hungry birds! Wherever you live, there is a feeder for you, with window feeders a great option if you have no outdoor space and a wonderful way to get close-up views. One thing to consider when choosing bird food and feeders is that different birds have different habits and favoured foods. Provide a mix of foods and you'll likely attract a greater variety of birds to your garden. Birds such as Robins, Blackbirds and Song Thrushes prefer to feed from the ground or from a bird table, and mealworms can prove particularly enticing. By contrast, tits, including Blue and Great Tits, and finches, such as Greenfinches, will be drawn in by hanging feeders and are happy to acrobatically peck at nuts, seeds, suet and fat balls. To tempt in that flashiest of finches, the Goldfinch, try sunflower or nyjer seeds.

Fights at the feeder

With food at such a premium, it's not surprising that squabbles and skirmishes are rife at the feeder. When it comes to battles at the bird table, size really does matter. Top of the flocks is the Great Spotted Woodpecker, one of the largest birds you'll see on your feeder and, while a thrill for us birdwatchers, one that's sure to send other smaller birds flying. At the other end of the scale is the Coal Tit. The smallest member of the tit family, it's often pushed off the feeder by larger Great Tits and Blue Tits. These latter two will stand their ground against each other but are in turn driven away by larger House Sparrows and Greenfinches.

There's also a pecking order within species, with males tending to dominate over the females, and juveniles faring even worse. Overcoming the rough and tumble at the feeder has led to some noticeable strategies for success. While larger species can afford to stay for longer at the feeder, smaller species will peck more quickly to make the most of their time there. Timing can also play a role. Feeders tend to be busier in the early morning and at dusk, so it's possible that some will choose to visit during the day when the competition is less intense. And there's always stealth. Watch out for a Robin dashing in to take a tasty treat, or look below the feeders for a Dunnock, picking up what's fallen down.

↓ You can help birds by putting out food.

Keep it clean

Not disinfecting your feeders regularly can lead to some very serious consequences. Between 1995 and 2020, Greenfinch numbers in the UK dropped by 68 per cent, and in 2021 the species was moved to the UK Red List, marking it out as in need of urgent conservation action. This crash in the population was down to a severe outbreak of trichomonosis, an infectious disease often spread through contaminated food and drinking water, and there are ongoing concerns that it could impact other species too.

To prevent the spread of disease, the RSPB recommends that you clean all bird feeders, tables and baths once a week using an animal-safe disinfectant. Wear rubber gloves and use a brush to scrub clean, removing any old food. If you do suspect that a bird visiting your feeder has been affected by disease, it's best to stop feeding for at least two weeks and

only start again when you are no longer seeing birds with any signs of disease. Birds with trichomonosis look generally ill, often lethargic, with finches often having matted plumage around their bills.

Water of life

Water too can be hard to come by in the winter, as some usual sources may freeze over. Birds need fresh water for drinking and some will also use it for bathing. Putting out fresh water in a bird bath is one way to help – and a largish plant saucer can be just as good. If the water does freeze over, you can add a little hot water to melt the ice. A more ingenious way to keep the water moving is to add a ping pong ball. As long as there is a little wind, the ball should bob about on the surface, preventing the ice from forming.

Keeping warm

While we can help our birds with food and water, we can't do anything about the cold. Birds, however, can. It's around this time of year that Robins often look their plumpest. But that's not a result of the food you've been putting out. Many birds fluff up their feathers and in doing so increase the amount of air held against their bodies. This helps prevent heat loss and keeps the bird warmer. Birds also grow more downy feathers in the colder months. These are the soft feathers found closest to their skin. It's a little like an extra layer of clothing to combat the cold!

↓ Starlings gather to bathe and drink.

Birds of the month

Robin

MEET THE GARDENER'S best friend, the winner of an informal poll which crowned them the UK's National Bird. Charming and fearless, Robins will frequently pop up as you garden, picking off worms and other invertebrates disturbed by your digging. This friendly nature has clearly earned them a place in the nation's hearts, but it's a clever tactic that they have long employed to find food. Like many of our garden birds, Robins were once primarily woodland inhabitants. There, they'd follow the snuffling and rooting of wild boar, but today, a human gardener is far easier to find. This willingness to 'work' with humans and a readiness to visit our feeders makes the Robin one of the UK's most familiar birds.

TOP ID TIPS

An adult Robin's rusty-red breast makes it easily identifiable. Male and female adults look alike. Juveniles have no bright bib but are marked out by their fine pale spots on a uniform warm brown background.

Unlike most other birds, Robins will sing throughout the year, making winter a great time to get to know their plaintive trills. Robins also break from the norm by being one of the few species in which females regularly sing. But these songs are not to brighten up our winter days. Robins are fiercely territorial and they sing to stake their claim.

WHEN AND WHERE TO SEE

All year round in gardens, woodlands and parks.

Great Tit

THE APTLY NAMED Great Tit is our largest member of the tit family and another woodland species that has happily moved into parks and gardens. Common throughout the UK, this dapper little bird is often one of the first to visit when you put up a nest box or install a bird feeder. It can also be one of the most vocal, with the male proudly declaring his residence with his familiar two-note song. During winter, Great Tits often form loose flocks with other small birds such as Long-tailed and Blue Tits. Head out on a countryside walk and keep your eyes open for these chirruping gangs roaming the land in search of food.

TOP ID TIPS

Great Tits have a smart glossy black head which extends into a neat line down the belly, set off by big white cheeks. Their blue wings and yellow bellies can see them mistaken for Blue Tits, but it's the black belly stripe that sets them apart. Sexes are similar, although the belly stripe is narrower and can be broken up in females. Juveniles are paler and drabber.

Although a Great Tit does have a fairly extensive variety of calls, the most easily recognisable is its repetitive see-saw song, often likened to *teacher, teacher, teacher*.

WHEN AND WHERE TO SEE

All year round in gardens, woodlands and parks.

Blue Tit

BLUE TITS are feisty little birds, who will fast take advantage of any help offered by us humans. Hang up a feeder or put up a nest box and it won't be long until a Blue Tit comes to call. These garden acrobats will happily hang upside down from feeders or fat balls as they peck away for food. You may also see them pecking at walls or window frames, a neat little trick to get at the spiders that lurk there. Blue Tits are also known for their unusual nest sites. While most tend to choose a hole in a tree or a nest box, nests have been found in bins, hollow road signs and even a drain!

TOP ID TIPS

Blue Tits are recognisable by their blue wings and yellow bellies. Unlike the larger Great Tit, their heads are topped with a blue 'beret'. Males and females are very similar. However, if you're lucky enough to see a pair, you may be able to tell them apart as the male's colours are noticeably brighter. Juveniles are drabber and yellower.

A Blue Tit's song can be difficult to get to know and does vary. However, Blue Tits will often begin with two high-pitched longer notes followed up with a rapid sequence of shorter calls, a bit like *tsee-tsee-di-di-di-di-di-di-di*.

WHEN AND WHERE TO SEE

All year round in gardens, woodlands and parks.

Long-tailed Tit

YOU MAY HEAR a Long-tailed Tit before you see it. These are gregarious little birds that hang out in family groups and call frequently to each other as they flit through gardens and woodlands. Spot one of these fluffy lollipops and it pays to keep looking, as there will likely be more around. In autumn and winter, groups can number as many as 20 birds. During the spring breeding season, adults who have failed to breed will help neighbouring parents feed their young. It's always a treat to see Long-tailed Tits as, although common, these restless birds are always on the move.

TOP ID TIPS

With tails longer than their bodies, it's easy to see how these birds were named. The body is a pom-pom of pale pink and white, while the tail and wings are mainly black. The face is white with black stripes above the eyes – in juveniles, the black is extended into a 'bandit mask' around the eyes. Males and females are alike. Despite their distinctive looks, they can sometimes be tricky to identify as they flit and fly, seemingly never still enough to get a good view!

Long-tailed Tits don't have a recognisable song like many other birds. Instead you'll hear them calling and chatting. Listen for a high-pitched *see-see-see* often broken up with a short *thrupp*, and chatty clicking calls.

WHEN AND WHERE TO SEE

All year round in gardens, woodlands and parks.

Coal Tit

A COAL TIT is a bit like a small, mainly black-and-white version of a Great Tit. Less commonly seen in our gardens, these small tits favour coniferous woodland. However, they will often visit feeders, particularly during the colder months when there are fewer insects and spiders around for them to eat. Their diminutive size means they come last in the feeder pecking order, so their visits to feeders are often fleeting to avoid the bigger birds. Coal Tits will also stash food, hiding it in various places in what's known as 'scatter hoarding'. They usually can't carry more than one seed at a time, so they will make several rapid trips with the intention of eating at leisure later.

TOP ID TIPS

The smallest tit found in the UK, the Coal Tit is compact, with a black cap, white cheeks, a white stripe on the nape, and a wide black bib under the bill. Their belly is pale buff, and they have olive-grey backs and wings. Males and females are alike, and juveniles are drabber with yellowish cheeks.

The easiest way to remember a Coal Tit's song is to think of a speeded-up, slurring Great Tit. Remember the Great Tit's *teacher, teacher, teacher* call? A Coal Tit's is a high-pitched *steecher, steecher, steecher*.

WHEN AND WHERE TO SEE

Mainly woodlands. Sometimes parks and gardens in autumn and winter.

January's challenge bird

Goldcrest

TAKING THE PRIZE for Europe's smallest bird is the tiny Goldcrest. At just 9cm from bill to tail, with a body not much bigger than a ping pong ball, its size can make it a challenging bird to see. Yet come winter and your luck might be in. The cold weather brings an influx of Goldcrests from Scandinavia, swelling our resident populations. They also tend to move down from the tops of conifer trees where they are usually found, foraging in the lower levels and branching out to other trees. They will sometimes join parties of Great and Blue Tits and other small birds.

TOP ID TIPS

A tiny greenish bird, with a round body and shortish tail. Goldcrests are named for the black-edged, bright stripe on their heads, which in males is orange-centred and in females is bright yellow. Despite the vivid colour, the central colour of these crests can be hard to see as they are only obvious when the bird is displaying or has fluffed up its feathers. You should get a glimpse in January as the birds tend to fluff up in the cold winter weather. Juveniles lack these head markings.

Somewhat in keeping with its tiny size, a Goldcrest has an extremely high-pitched call and, if you're lucky enough to have good hearing, its sharp *see, see, see* can alert you to its presence.

WHEN AND WHERE TO SEE

All year, particularly in mature coniferous woodland in parks and gardens.

Monthly musings

History of bird feeding

These days, about half of all householders in the UK feed birds, but where did the idea come from? Casual bird feeding has been around for millennia and is mentioned in Hindu writings from 1500–500 BC. In the UK, one of the earliest records is of St Serf of Fife who fed a tame Robin back in the sixth century. Bird tables like those we have today were used from the 1850s. A famous German bird book about how to feed birds was published in 1901, and by 1910 there was an article in *Punch* magazine about how popular it was becoming. Tubular feeders came in 1960, which kept food dry. More recently, the idea of feeding birds all year round became accepted and popular.

Daring Dippers

The Dipper is one of our most remarkable birds, making a living by diving into rushing water for insects and their larvae, especially mayflies, stoneflies and caddisflies. They are the world's only aquatic songbirds, and they perform many marvellous tricks, such as diving into the water in mid-flight, and 'walking' underwater on the river bottom, against the current. They are also incredibly hardy. They sing in midwinter and in northern areas have been seen feeding under ice, as well as in glacial rivers. Another of their impressive quirks is that they sometimes nest behind waterfalls. What a bird!

↓ A Dipper with a catch.

↑ Clockwise from left to right: Great Spotted Woodpecker, Green Woodpecker, Lesser Spotted Woodpecker.

Top two must-dos

1 Seek out the woodland drummers

January is a great time to listen out for the explosive drumming sound of Great Spotted Woodpeckers as they take to the trees to make their presence known.

The drumming is purely for show, to mark their territory and communicate. It's not the sound of them digging out a nest hole or pecking for grubs. Both male and female woodpeckers drum and they will often return to trees with the right kind of sound. They are looking for a sound that resonates through the forest, so favour hard dead branches, but they have also been seen drumming on telegraph poles and even drainpipes.

It's an incredibly intense drum, with Great Spotted Woodpeckers packing in 10 to 40 strikes per second! Until 2022, it was thought that a woodpecker's skull was adapted with a frontal spongy bone that acts as a shock absorber to protect the brain. However, new research disputes this, showing that their heads act more like stiff hammers rather than shock absorbers. This suggests it's simply their smaller size that protects them. While us humans would suffer severe concussion, woodpeckers can happily hammer away.

For the best chance of hearing and seeing this rather remarkable spectacle, head to the woods in late winter through to early spring and listen. Great Spotted Woodpeckers can be found in mature woodland throughout Britain, as well as parks and gardens.

Look out for a striking black-and-white bird. Great Spotted Woodpeckers are Blackbird-sized, with black backs and wings and white shoulder patches. Their belly is mainly white, with a deep red patch under the tail. You can tell males and females apart by the male's red patch at the back of the head.

Britain's native woodpeckers

Britain has three native woodpeckers: Green, Great Spotted and Lesser Spotted. Green Woodpeckers have olive green wings and a red cap. Unlike Great Spotted and Lesser Spotted Woodpeckers, Green Woodpeckers don't often 'drum'. Instead they make a 'yaffling' or laughing call. You're also more likely to see them on the ground, feasting on their favourite meal of ants, rather than on a tree.

Great Spotted Woodpeckers are much larger than the diminutive Lesser Spotted Woodpecker, which is roughly the size of a sparrow. They are also much more common. Lesser Spotted Woodpeckers are scarce birds, with only a couple of thousand pairs found in the lowland woodlands of England and Wales.

2 Visit a nature reserve

January is a great time to enjoy nature. Bare branches can make it easier to see woodland birds, wetlands can be brim-full of wintering waterbirds such as ducks and geese, and bird feeders on nature reserves are likely to be packed with hungry birds!

Chase away the January blues

Getting out and about in nature can also do wonders for our health and wellbeing. It seems the scents, sights, feel and sounds of nature can stimulate our senses, helping to put our minds at rest and our bodies at ease. Birdsong in particular has been found to improve mental wellbeing. So open your window and listen to the birds. Or head outside and immerse yourself in nature.

Discover a wonderland of wildlife

The UK's landscape is incredibly varied, from coastal cliffs and sandy beaches to rolling countryside and ancient woodland. And this makes for a huge range of different habitats and nature experiences, right on our doorstep.

Go west and you could enjoy one of the rarest habitats on earth, the UK's temperate rainforest. These wet woodlands, such as RSPB Carngafallt in Wales, the Woodland Trust's Crinan Wood in Scotland or Glengarriff Woods nature reserve in Ireland, are *Lord of the Rings*-like wonders, with mature trees dripping with mosses and lichens. Take a trip to the east coast and you could enjoy birdwatching on expansive estuaries such as RSPB Blacktoft Sands on the Humber or the Thameside Nature Discovery Park, managed by Essex Wildlife Trust.

Head for the hills and discover the mountain wonderlands of the Cairngorms National Park in Scotland, the Mourne Mountains in Northern Ireland or Eryri National Park in Wales. Or why not wander to a wetland paradise? Reservoirs, lakes, ponds and water-filled gravel pits are often fantastic places to see birds, particularly in winter when ducks, geese and other wetland birds flock to them in huge numbers. Try Rutland Water in the Midlands, WWT's London Wetland Centre or RSPB Lake Vyrnwy nature reserve in Wales.

There are thousands of places managed for wildlife across the UK, with many of them free to visit, and there's sure to be one near you.

CHOOSE YOUR OWN ADVENTURE

National Trust
The National Trust manages more than 500 places including a number of National Nature Reserves. nationaltrust.org.uk

The RSPB
The RSPB has a network of more than 170 nature reserves across the UK. rspb.org.uk

Wildfowl & Wetlands Trust
WWT is the UK's leading wetland conservation charity, with 10 wetlands centres. wwt.org.uk

The Wildlife Trusts
There are 46 local wildlife trusts, with 2,300 nature reserves across the UK. wildlifetrusts.org

Woodland Trust
Find a woodland near you by visiting the Woodland Trust's website. woodlandtrust.org.uk

How to help

Do the Big Garden Birdwatch

Join in with the RSPB's Big Garden Birdwatch and help monitor how garden birds are faring. Big Garden Birdwatch began back in 1979 as an activity for children that was promoted on BBC's *Blue Peter*. It soon caught on, and it's now the world's largest citizen science wildlife survey, with around half a million people across the UK taking part annually.

The RSPB calculates that more than 190 million birds have been counted since Big Garden Birdwatch began, and more than 12 million hours have been spent recording garden birds.

Held over one weekend in late January, all you need to do is spend one hour counting the birds you see in your garden, from your balcony or in your local park. It's a great excuse to put your feet up and do nothing but watch nature.

January's a prime time for garden birdwatching, as the colder weather and the lack of natural food brings many birds to our feeders and bird tables. Birds can take a little while to start using a feeder, so if you're hoping for bumper numbers in your Birdwatch, it's a good idea to start feeding earlier in the month. It's also a great way to get to know the types of birds that visit. Start with a simple hanging feeder and watch them flock in!

↑ House Sparrow in the snow.

Counting for nature

The real beauty of the Birdwatch is the data generated by more than 40 years of surveys. Big Garden Birdwatch was one of the first wildlife surveys to highlight declines in Song Thrushes and Greenfinches. First ranked at number 10 in the Birdwatch in 1979, sightings of Song Thrushes have steadily declined over time, dropping by a whopping 85 per cent. Greenfinches have similarly dropped out of the top ten, with numbers sighted down 69 per cent since the Birdwatch began. Findings like these can help the RSPB draw attention to what's happening in the bird world and better steer the efforts of conservationists working to safeguard nature.

The RSPB's Big Garden Birdwatch is also picking up on some recent changes thought to be down to climate change. Blackcaps are pretty warblers that migrate from southern Europe and North Africa to nest and raise their chicks in the UK every summer. However, in recent decades increasing numbers of Blackcaps have been sighted during the Birdwatch, with ringing studies showing that these wintering birds are mostly visiting us from relatively colder eastern Europe.

Birdwatch number one

In 2024, the House Sparrow celebrated its 21st time as top of the flocks, being the most commonly sighted bird in the Big Garden Birdwatch since 1993. Despite being number one, House Sparrow counts have actually dropped significantly, down by 60 per cent compared to the first Birdwatch in 1979. This change underlines the general decline in many of our birds, with an estimated loss of 38 million birds from UK skies in the last 50 years. It's stats like these that make doing January's Big Garden Birdwatch more important than ever.

To take part, simply visit the RSPB's website, where you'll find everything you need.

↑ Birds such as Blackcaps can be enticed by apples.

Myth of the month

Magpies: malevolent or magnificent?

My aunt always insisted on saluting a Magpie when she saw one, to be sure to avoid bad luck. While my no-nonsense dad scoffed at these actions, my aunt was far from alone in her superstitions. Magpies, perhaps more than any other common bird, are associated with bad luck, with the famous rhyme warning of 'one for sorrow'. One theory offered is that Magpies, along with that big black bird of bad omen, the Raven, would often turn up in places associated with death, such as battlefields or gallows sites. But this doesn't explain why the rhyme moves on to suggest 'two for joy'. Surely two scavenging Magpies is even more macabre than one!

Whatever the origins of the rhyme, Magpies retain a malevolent reputation, often chased from bird tables and gardens with the aim of protecting smaller, more popular birds. One more recent myth is that an increase in Magpies has driven a decrease in songbirds. However, research by the British Trust for Ornithology showed no evidence for this, finding that songbird numbers were no different in places where there were many Magpies to where there were few.

What is true is that the Magpie deserves a closer look. Rather than salute it, spend a little time admiring its 'black' feathers and you'll see a shimmering iridescence, especially on its magnificently long tail feathers. This is surely the stuff of fairy tales.

← Spreading bad luck or good tidings?

FEBRUARY | 2

It takes two

The cold, the short days and the bare branches of
February conceal a world that, for birds, is radically changing.
The breeding season is already steaming ahead.

IN FEBRUARY IT CAN FEEL as though winter will never end, but if you look carefully, spring is accelerating into view. There are signs everywhere of an increased friskiness; the garden Blue Tits are definitely in pairs, the dawn chorus is louder and more varied with every passing week, and some birds are nest-building. Rookeries surge with sound. Pairs of Rooks are refurbishing their structures in the bare treetops, and it is as noisy as any building site.

The emerging spring is a physiological reality for birds. Their preparations for breeding are already well underway and have been ever since the winter solstice in late December. Amazing though it might sound, birds can detect light deep in their brain – it passes through their skulls – and they are sensitive to increasing day length. The increase of light is the only reliable, unchanging trigger they have to set a cascade of hormones into action, so that they reach full

→ Greenfinches on your feeders in February have more on their minds than just survival.

← When Mute Swans perform their elegant courtship display, their necks make a heart shape.

breeding condition at the optimum time. Temperature and weather are far too fickle. Another remarkable biological fact is that after breeding, in late summer, most birds' ovaries and testes degenerate to a fraction of their size – when they are not needed, their extra weight would only be a burden to a bird that needs to fly. In spring, this has to be reversed, and it is one of many physiological and behavioural changes that are triggered by the effect of longer days on the avian hormonal system.

We cannot see any of these biological changes taking place within birds' bodies, but there are plenty of outward signs, the most obvious of which is song (see April, page 59). The outdoors is getting noisier and noisier. Even in January, almost all the resident birds, including Blue Tits, Great Tits, Song Thrushes, Dunnocks, Wrens and Robins, are already in full song; Blackbirds and Chaffinches start in February. Already, many individuals are spending hours in song a day, despite the adverse conditions.

Courtship

You'll see plenty of early evidence that birds are preparing for the breeding season if you watch them carefully enough. Note how the Greenfinches are in pairs, sitting side by side on feeders. You might see Blue Tits examining nest boxes as a duo. If you are fortunate enough to spot one Treecreeper, you will soon see another. And you'll notice more easily that Collared Doves, Magpies, Carrion Crows and House Sparrows spend much of their lives in pairs. Pigeons and doves, for example, may sit side by side, preening each other. The tone is changing.

↓ Yellowhammers usually find their mates in winter flocks.

It is still winter, of course, and many bird species are still mostly seen in flocks, large or small. However, the dynamic of the flocks is changing. If you think about it, what could be a better place for a bird to meet a partner? Birds such as Siskins, Blue Tits, Yellowhammers, Lapwings and Starlings have the perfect chance to assess the quality of their peers in winter flocks, and the same happens among gulls and pigeons. It is perhaps, though, most obvious among wildfowl. Ducks have been displaying since the autumn and now you can easily see species such as Mallard, Gadwall and Shoveler in pairs and taking up their own small portion of a lake or pond in a display of twosome-ness. When migratory ducks pair up, the male accompanies the female to her birthplace, where they breed.

↓ Starlings often flap their wings when they are singing.

Even now, in February, you can enjoy displays of all kinds: some are purely territorial, the bird establishing its own patch of habitat with a view to breeding; some are direct courtship towards an already interested or established mate; others are a bit of both. All are good entertainment to us. Already, even so early, the Lapwings are making cartwheels in the air, flying as if all shackles were off, and whooping as if wildly drunk. Skylarks take to the eponymous sky and throw down song as if they were billionaires donating money. Blue Tits make a subtle flight from one branch to another, using super-fast, bat-like wingbeats. Greenfinches take off from the top of a tree and flit in graceful circles at treetop-level. Starlings perform peculiar wing-flaps while perched, Jackdaws tumble in the sky, Mute Swans swim towards one another to make a heart shape with their necks, Dunnocks wave their wings, gulls nod towards each other and Kestrels fly high in the sky, allowing the pale undersides of their wings to shimmer in the weak February sun. It's hot out there in the cold!

Birds display whatever their situation, whether they are long-established members of a couple, are re-pairing after the loss of a mate, or facing their first experience of hormonal excitement. Performing effective visual displays is just as essential to breeding as generating all those hormones. By making the right moves at the right time, a couple can coordinate their body chemistry and take steps towards successful breeding.

Complicated love lives

Birds' relationships are often more complex than we realise. Many birds veer from monogamy in one way or another. For example, in birds as diverse as Wrens, Cetti's Warblers, Corn Buntings and Black Grouse, the most desirable males frequently pair up with several females. The best-looking male Pheasants hold harems. Dunnocks have a famously complicated mating system in which both males and females routinely form significant pair bonds with several of the opposite sex. Moorhen females sometimes pair up with their fathers. And it is probably the rule, rather than the exception, that males or females in established pair bonds will copulate on a 'no strings' basis with others, should the opportunity arise. In some species, such as Reed Buntings and Coal Tits, this happens to an extraordinary extent. About one in three Coal Tit chicks are sired by a neighbouring male, not the one that feeds them.

Making tracks

Of course, there is one more highly significant event happening in February that we don't really notice at all because it is happening far away. But, just as our resident birds are already geared up for breeding, the same is true of the birds that come here for the summer. Many of them, such as Swallows, have already left their winter quarters and are well on their way. On a day in February in the lingering cold, that is another prospect to warm the heart.

↑ Male Dunnocks regularly compete to mate with females. This 'wing-waving' display is made by a male towards a rival.

Birds of the month

Chaffinch

THIS IS A BRIGHT, breezy bird, with colourful plumage and a cheery song to brighten any winter day. At this time of the year, it is highly sociable, and can be found in woodlands, field edges and gardens. You can often disturb flocks feeding on the ground, flushing them up with a flash of white wing-bars. At the moment they primarily feed on seeds, often collecting fallen mast in Beech woods, but they also readily come to feeders. Later in the year, their diet will switch completely to insects, which they also feed to their young. In February, the song is just beginning. It is a cheerful, accelerating rattle with a flourish at the end, and you'll hear it in most woodlands. Some Chaffinches seem to be bad singers, leaving off the flourish at the end; these are young males that are still perfecting their song.

TOP ID TIPS

A slender finch with a long tail with white outer margins. It often feeds on the ground and has a curious waddle, keeping its head still as it walks, like a chicken. It has white shoulders and a complex set of white markings on the wings. The male is colourful, with a blue cap, pink breast (the call is also *pink-pink!*) and chestnut back. The female and juvenile are much more monotone fawn-brown.

WHEN AND WHERE TO SEE

Abundant in woods (of all kinds), fields and gardens throughout the year.

↓ In early spring, the male Chaffinch (left) acquires its colourful plumage as dull feather tips wear off to reveal brighter colour beneath. The female (right) remains dull in colour.

Greenfinch

MOST PEOPLE KNOW the Greenfinch best from their garden, where it is a regular visitor to feeders. It is a heavily built finch with a broad bill, and it has a slightly bullying temperament, not allowing other birds near it when it is monopolising food. The large bill allows it to crack open a wide variety of seeds, from minuscule weed seeds on the ground to sunflower seeds on flowerheads. It is also fond of yew in the autumn. At this time of the year, it is especially noisy, singing atop tall trees (often cypresses), bushes and even aerials. Listen out for the trilling song, which often incorporates a long, drawn out 'wheeze'. It has an easily missed display-flight, in which it takes off from a high point and flies in a wide circle or figure of eight, flying with deep wingbeats and pitching from side to side – it can look like an early Swallow. Greenfinches are highly sociable birds and often go around together in flocks.

TOP ID TIPS

The build resembles a sparrow and the two can be confused, especially the female Greenfinch (see photo on page 23), which is duller than the apple-green male. The juvenile is like a streaked version of the female, but all Greenfinches always show a yellow wing-bar and yellow sides to the tail. The big pale pink bill is also distinctive. Its flight has an up-and-down, bounding motion. The main call is a soft *chip-chip*.

WHEN AND WHERE TO SEE

Common in gardens, but also in churchyards, scrub, woodland edges and hedgerows, throughout the year. It has suffered from an outbreak of the infectious disease trichomonosis in recent years, which means it is now quite rare in some areas.

Goldfinch

IF THE GOLDFINCH was on social media, you might be suspicious of its perfect life. It is stunningly attractive, has a delightful song, is sociable and lively and even builds a beautifully constructed nest. Most people get to know Goldfinches when the birds visit feeding stations, usually in small groups which all like to feed together, using all available perches. They come at any time of year. Away from the garden, Goldfinches can be found wherever there are seeding plants, especially thistles. They have a habit of feeding on the seedheads of plants like thistles and teasels, which are often unstable, so they need to flutter their glittering wings to keep their balance. When feeding, they insert their thin bill between the bracts and prise them open to reach the seeds.

TOP ID TIPS

It is unmistakable if seen well. The bold red, white and black face pattern, together with the brilliant yellow wing-bars, add colour to the drabbest of winter days. Juveniles, which you'll see in summer, have plain brown heads and a forlorn expression. It makes a variety of effervescent twittering calls and its flight note is *tickle-it*.

WHEN AND WHERE TO SEE

A very common bird in gardens, parks, hedgerows and overgrown, weedy places throughout the UK, except for uplands.

↓ An adult (right) and juvenile (left) Goldfinch investigate a teasel head.

Bullfinch

THE BULLFINCH clearly gets invited to all the posh parties – few British birds are so perfectly turned out in top-and-tail smartness. The startling plumage, with its neat contrasts and vivid tones, sets this finch apart, even amidst its colourful tribe. It always feels like a privilege to see one, because this is a shy and retiring introvert, not at all noisy and usually only seen in small, dinner party-sized groups. Compared to other finches it eats far more soft fruit and buds, as well as seeds. It can eat up to 30 tree buds in a minute. In spring, the adults develop small pouches in the floor of their mouths to store food before they bring it to their young, with a total capacity of a cubic centimetre.

↑ Bullfinches (male left, female right) are found in pairs and family parties all year.

TOP ID TIPS

The smart black cap, found in both sexes, instantly separates it from the similar Chaffinch. Juveniles don't have black caps, but plain brownish heads with very beady-looking eyes. The bill is short and broad and makes a neat continuous curve with the forehead. Its plumage looks soft, and the intense pinkish-red colour of the male's breast is unmistakable. Often all you see is a bird flying away, showing off another signature feature, the dazzling white rump. The call is a soft and slightly melancholy *pew*.

WHEN AND WHERE TO SEE

Fairly common throughout the country all year in scrub, woodland edges and hedgerows. It sometimes visits gardens in winter, including feeding stations.

Siskin

THE SISKIN is a waif of the treetops, and most people would probably never see it if it wasn't for its February resolution – to find a free supply of seeds in gardens. Earlier in winter, it has been a lakeside and riverside bird, munching away at the ripening seeds of Alder trees, and before that it would have been taking advantage of birch seeds from July to October. Siskins feed in large flocks (often 30–100), and typically 'burst' out of the branches if they are disturbed. Now the Alder seeds are all but spent, they need to forage elsewhere, so this is the best time to see this lovely bird in the garden. Soon it will migrate north and breed in spruce forests, where the cones are ripening.

TOP ID TIPS

It looks a bit like a Greenfinch that's been on a dramatic weight-loss programme – so it is Blue Tit-sized rather than sparrow-sized, with a much smaller, thinner bill. It has yellow bands across its wings, rather than the Greenfinch's stripe along the edge. And it also has strong dark streaking, especially on the flanks. Males have a coal-black chin and crown; females and juveniles have plainer heads. The call is a nasal and creaky *dzwee!*

WHEN AND WHERE TO SEE

Most easily seen at feeders in gardens, and the end of winter is the best time. Otherwise, you may see Siskins in larger conifer woods and, in autumn and early winter, in Alder trees. Widespread, but they are mostly a winter visitor to lowland England.

↓ The male Siskin (top) is more brightly coloured and has more head markings than the female (bottom).

February's challenge bird

Hawfinch

THERE ARE PEOPLE who think that the Hawfinch is a figment of the imagination. It isn't, of course, but it is a challenge to find. It is famously shy, flying fast and far when disturbed. It has an annoying habit of perching not quite on a treetop, but just below, hidden in the upper branches. And it is quiet. It has no chatty song like a typical finch, and its spitting *tsip* call is easily missed. The sounds it makes are never loud.

The Hawfinch's claim to fame is its remarkably broad bill which, along with thick neck muscles, enables it to crack hard seeds such as those of cherries, Hornbeams and, on the continent, olives. It has bumps on the floor and roof of its mouth, which help it to secure larger seeds as it splits them in half with a mighty bite. Remarkably, the bill has a biting force of 50kg! This power enables Hawfinches to exploit seeds unavailable to other birds, as well as any other seeds it wishes.

But it isn't a one-trick pony. In the summer, Hawfinches eat some insects and can even catch them in flight.

TOP ID TIPS

A large, chunky finch, with its signature massive bill, oversized head and a noticeably short, white-tipped tail. The overall colour is similar to a Chaffinch (with juveniles rather plain brown), but note the plump shape, black chin and broad pale wing-bar.

WHEN AND WHERE TO SEE

Found mainly in large, deciduous forests during the day, although it often comes to roost in large, thick conifers. There are spots throughout Britain where birds can be seen congregating in the evening. It is widespread, more so in winter.

↓ The Hawfinch is famously elusive in the treetops.

Monthly musings

The growing Red List

The Birds of Conservation Concern Red List is an assessment of the conservation status of UK birds, compiled by bird conservation and monitoring organisations including the British Trust for Ornithology and the RSPB. Birds on the Red List are those identified as being most vulnerable, having suffered huge declines or only existing in vulnerably small numbers. The number of birds on the latest Red List (published December 2021) is 70, a huge increase from 36 when the first assessment was made back in 1996. Newer additions include once familiar garden species such as House Sparrow, Greenfinch and Starling. It's fair to say that our birds really could do with all the help we can offer.

Back away from the bush!

The scarcity of food in winter can lead to some very territorial behaviour. Mistle Thrushes are Europe's largest thrush, so named for their traditional diet of Mistletoe berries in winter. But given the wide variety of berries they eat, perhaps 'Berry Thrush' would be a better name. In fact, Mistle Thrushes are so enamoured with berries that they will often take ownership of an entire berry tree, fiercely defending it from others looking for a bite to eat. You may see one on guard, watching from a high vantage point, sometimes singing to mark its territory. But woe betide a hungry visitor, as the Mistle Thrush will soon chase away any intruders and call angrily with a loud rattle or churring.

↑ Mistle Thrushes may defend clumps of berries, such as holly, throughout the winter.

Top two must-dos

1 Watch a Barn Owl

One of Britain's most charismatic birds, the Barn Owl is quite tricky to see. But with the nights still long and starting early, February is a good month to see one. You need to find a rural location with plenty of long grass – often a meadow or rough grassland – and go before the light fades. Select a dry evening without too much wind; owls don't like flying in the rain. Wait. And expect a thrill.

There is nothing quite like an encounter with this bird. The Barn Owl just seems to materialise in the gloaming, rather than make an entrance. The plumage can look as white as a ghost, but if the light is good, you will notice the gorgeous dappling of ochre, grey and rich honey on the upperparts. The owl's flight, on rounded wings, is light, floating and wavering, low over the ground. It often follows a line, such as a ditch, but then suddenly goes off at a productive-looking tangent. Progress is interrupted by brief hovers and, if prey is spotted, a quick drop, talons first. The hunt of the Barn Owl is so mesmerising to watch that you forget it culminates in a death, usually of a vole, mouse or rat.

↓ The Barn Owl's flight is buoyant, wavering and perfectly silent.

Silent hunters of the twilight

The extraordinary thing about all owls is the absolute silence of their flight. The flight feathers are large for the size of bird and extremely soft. The leading (narrow) edges of the feathers are fitted with a comb-like serrated fringe which minimises turbulence and reduces flapping noise. No human can hear them coming and, of course, the idea is that prey doesn't either, so that it is taken by surprise. Don't forget that the owl needs to be able to hear the sound of squeaking or rustling mice over the sound of its own wingbeats, too.

Barn Owls are famous for their ability to catch food in complete darkness, and they achieve this by having three-dimensional sound detection. The left ear is placed lower on the skull than the right ear, meaning that it can pinpoint the exact position of moving prey, even directly below, and make the perfect pounce into the long grass. The heart-shaped facial ruff also channels sound towards the ears. The long legs are perfect for grabbing prey through the grass, and the talons are suitably sharp and lethal. Not only are these birds gorgeous, but they are also highly adapted hunting machines.

2 Marvel at a Great Crested Grebe courtship

Ballet is coming to a lake near you. The long breeding season of the Great Crested Grebe is beginning, and pairs of these elegant birds have already embarked on their courtship, which is one of the most eye-catching of any in the UK. If you take a walk around a lake at this time of year, you could see something amazing. Listen out for the

↑ Head-Shaking, with the birds facing each other, is an important part of Great Crested Grebe courtship.

peculiar hollow braying calls of the birds, or their rapid cackles, which often signal displaying afoot.

Head-shaking

The commonest grebe ceremony is known as Head-Shaking and is easy to see and recognise. The two birds face each other, ruffle their colourful head-plumes and make loud calls, mainly a sort of cackling quack. While doing so, they shake their head from side to side – in Great Crested Grebes, a shake of the head definitely means yes! – sometimes gently, sometimes in bursts. Every so often, they will also bend down to rest the head on the back feathers, and even to touch a feather in a mock-preening action known as Habit-Preening.

35

Other ceremonies

These birds have a complex series of ceremonies, and these are of particular interest because the sexes swap roles, with no posture or movement confined to one or the other. The most exciting display is known as the Weed-Dance, and it is a dramatic must-see. In this ceremony, the birds interrupt Head-Shaking with an abrupt dive underwater. If they come to the surface with waterweed in their bills, keep watching! After a moment, they will swim together, each holding the weed and, as they meet, rise up into a vertical posture with their feet furiously paddling to keep them upright, breast-to-breast. As they maintain this so-called Penguin Display, they wag their heads from side to side, the weed flapping in their bills.

Unsurprisingly, given the effort expended, the Weed-Dance is brief. But seeing it is a golden moment in early spring.

Another common routine seen in the early part of the season is called the Discovery Ceremony. One bird of either sex dives down and approaches the other one underwater, in the so-called Ripple Approach. It then rears up half out of the water, while the watching bird ruffles its feathers. There is also a Retreat Ceremony, when one bird suddenly stops Head-Shaking and flies away over the water, its feet pattering wildly and making a splash. The other bird watches, ruffles its feathers and then the pair drifts quietly back together.

↓ The most thrilling of the Great Crested Grebe's displays is the Weed-Dance.

How to help

National Nest Box Week

Many birds will be breeding soon, so what better time to think about giving them a helping hand? National Nest Box Week has been run in the third week of February since its inception in 1997. The fact it starts around Valentine's Day means that you can celebrate love by helping out a feathered couple.

The main thrust of the Week is to encourage you to make a nest box, although you can validly salute it simply by putting up a shop-bought one. Making one is easy, good fun and well worth the trouble; why not join one of the dozens of special events run by the RSPB and other organisations, where they will provide you with the materials and instruction you need? The whole family can get involved and it teaches children that they can make a genuine difference to the natural world.

↑ A Blue Tit investigates a woodcrete nest box.

That's the joy of the whole thing. Nest boxes are not faddish accoutrements that make you look good but leave the birds cold. They are genuinely useful and many species, from tits to flycatchers, use them, sometimes more than 'natural' holes. In the wild, tree holes and other cavities are always in demand, with a long waiting list of hopefuls. Your provision will help a family in need. And if you have made the nest box yourself, the delight gained by seeing a Blue Tit family fledge, for example, will be genuinely personal.

There are many different types of nest box you could put up. In gardens, tits prefer hole-fronted nest boxes and Robins will use the open-fronted designs. Birds such as Swifts (see pages 81 and 110) benefit from specific designs. For most

↑ Kestrels benefit enormously from nest boxes, as breeding sites are always at a premium.

boxes, wood is a great material but must be treated with water-based, animal-safe varnish to prevent it decaying too quickly. Woodcrete boxes are great and last longer. It is good practice to clean out a box in the late autumn, so make sure it has a removable lid.

The birds you attract will depend on what is already around and what your local environment is like. For example, if there are no conifers in the immediate vicinity, you are highly unlikely to get breeding Coal Tits. Equally, if you don't have sparrows anywhere nearby, putting up a series of boxes for them will be, at best, a test of patience. It's worth trying anything and shooting for the moon, but make sure you are realistic in your expectations. Most gardens make do with Blue Tits and Great Tits, which provide great entertainment.

There is much hot air spoken about where you should site a nest box, but a good rule is that it is better to be out there than kept in storage indoors. A hole-fronted box can go on a tree trunk or a wall for tits; open-fronted nest boxes are best inside cover such as Ivy, to protect their occupants from passing predators. There is some merit in making sure it doesn't face due south (to avoid overheating from the sun), but other than that, the birds' opinions are the ones that matter.

Finally, be ambitious. Why not get permission to put up boxes in parks, workplaces and amenities, such as playgrounds? The sky is the limit.

Myth of the month

Birds and Valentine's Day

Ever since medieval times, people have associated St Valentine's Day (February 14th) not just with human romantic love, but also with the pairing of wild birds. Geoffrey Chaucer (1345–1400) wrote in his *Parlement of Foules*:

> *'For this was sent on Seynt*
> *   Valentyne's day*
> *Whan every foul cometh ther*
> *   to choose his mate.'*

The idea that birds pair up in mid-February is intuitive – after all, most people who love the outdoors will notice a burgeoning of song at this time of the year. The whole countryside is awakening, and soon the breeding season will be in full swing.

For many species February is pairing-up time, but it often happens earlier. Rooks and sparrows pair up in autumn, and Robins often have a significant other by December. Birds such as Magpies, geese, gulls and Woodpigeons pair for life anyway, so any display is intended to rekindle things rather than spark them. Many others, such as ducks, Siskins and Blue Tits meet their significant other during the winter in flocks. So if anything, most birds really should have their mate by now, and if not should perhaps be redoubling their efforts.

But as bird myths go, this one has a lot of truth to it.

↓ Mutual preening is a common feature of pigeon courtship.

Nesting time

Keep an eye on your garden birds this month and you could be rewarded with witnessing some very exciting behaviour as birds set about preparing to welcome the next generation.

→ Blue Tits and other birds may use wool to line their nests.

← A Wren peers out from its nest hole.

SIT QUIETLY in a park or garden in March and you may see some curious behaviour. Is that a Long-tailed Tit tugging at a spider web? Why is that Song Thrush pecking at mud? Then, as a Blackbird with a small twig in its bill gives the game away, the penny drops. It's peak nest-building time.

Timing is everything

Nest-building is part of a carefully timed series of events, designed so that the chicks hatch when there is likely to be good food availability. Among the first to nest are Rooks. These sociable crows can start as early as December or January, building their nests high up in trees in large groups. Worms are the favoured food of choice for their chicks, so they nest and breed early to avoid the drier summer months, when worms bury themselves deep in the soil out of reach. Similarly, Blue Tits build their nests in late March so that the hatching of their chicks coincides with the emergence of tiny caterpillars.

Although birds such as Rooks and Carrion Crows nest high up in trees, the reality for the majority of birds is very different. Most small garden birds, such as Robins and Wrens, choose to nest within a few metres of the ground, in hedges or shrubs, where there is a dense coverage of leaves and greenery. At this time of year, you may well see birds darting to and from the hedgerow. If you do, it's a sure sign birds are nesting as they ferry back and forth with nesting materials and then later in the season dash back and forth with food for their chicks.

As a general guide, the RSPB recommends not cutting your hedges between March and September to avoid disturbing nesting birds.

High spec homes

The materials birds use to build their nests can be surprising; it's not just twigs and feathers. One of the most intricate and beautiful nests is that of the Long-tailed Tit.

The nest begins with a pair dropping moss on a branch. The birds then use spider web to secure the mossy platform. Using more moss and spider web, they work on the sides until they have a cup shape. Next, it's time for a bit of lichen pebbledash to camouflage and strengthen the outside, fixed in place with more spider web. The birds then repeat the process to build the sides further and add a roof. Finally for the home furnishings, a luxuriously soft inner cushioning of

up to 2,000 feathers. It's a process that takes around three weeks, but it's built to last. Thanks to the spider silk's natural elasticity, the nest can expand as the chicks (and their tails!) grow.

Mud is another popular material among garden birds. Blackbirds make a nest using straw, twigs and other plant materials which they then line with mud and a layer of soft grass. Song Thrushes also line their nests with mud but forgo the grass. They will also add other materials to their dry mud lining, such as clay and even cow dung – although perhaps not in the average garden!

A Blackbird's nest is the epitome of what we might expect a nest to be. They are round and cosy-looking, with the twigs, grasses and moss all woven together. Not all nests are so

↓ Long-tailed Tits use lichen to clad their nests.

→ Collared Doves build notoriously ramshackle nests.

meticulously crafted, however. One of the most unkempt and messiest nests is undoubtedly that of the Woodpigeon. This bird simply drops large sticks and twigs in the fork of a tree and hopes for the best! Collared Doves too are notoriously slap-dash nesters, making a similarly precarious sparse platform of twigs.

Sharing the chores

There is no straightforward division of the sexes when it comes to nest-building, with different species employing different strategies. While Long-tailed Tit males and females work together, the Blackbird female takes sole responsibility for her nest. As she builds, she wiggles within the nest, moulding it to the shape of her body. Blue Tit females also work alone,

building a cup-shaped nest out of moss and feathers with little or no help from the males.

For Wrens, nest-building is part of courtship, with the male taking the lead. A male Wren makes several dome-like nests out of moss, grass and dead leaves, enticing a female to take a look, serenading her as he shows off his craftsmanship. Should she approve of his work, they will mate and for now that's pretty much the end of the relationship. Now it's over to the female to finish the nest, by lining it with feathers or fine hair. Then she begins the hard work of laying, brooding and feeding the chicks until they fledge.

Once the chicks have fledged, the male may step in again briefly to feed them. But it's likely he's been busy too. A male

Wren can build up to 10 nests, and if he can find another one or two females who like a nest, he'll mate with them too!

Looking after the eggs

Different birds have different strategies when it comes to looking after the eggs, too. Generally, however, it's the female who takes charge of incubation. This involves a good deal of care, with birds frequently turning the eggs to make sure that every part of each egg is warmed. Female Wrens do this alone, taking a break every now and then to forage for food. Female Blue Tits and Great Tits also incubate their eggs alone, but here the males will drop in with food for the females while they are on the nest.

For other birds, it's a more equal division of labour. Among House Sparrows, Starlings, Woodpigeons and Collared Doves, incubation is shared between the male and female birds. Much like in the human world, there's no 'one size fits all' when it comes to sharing household duties!

↑ A female Blackbird broods her eggs.

Could there be trouble ahead?

There is evidence that climate change is having an impact on when some birds nest. The British Trust for Ornithology's BirdTrends 2022 report identified a number of birds breeding earlier, with Greenfinches now laying their eggs 22 days earlier than in the 1960s. Typically, chicks hatch when the food they rely on is in peak availability. However, when birds breed earlier in response to climate change, there is a concern that there may be a mismatch between the chicks hatching and their favoured foodstuff being available. As yet, the jury is out on how much impact this will have on the breeding success rates of different bird species and much more research is needed. Nature's Calendar, hosted by the Woodland Trust, is the UK's largest phenology database, recording the timing of natural events, including when birds nest and chicks fledge. You can add your own sightings and records to help scientists investigate the effects of weather and climate on wildlife.

Birds of the month

House Sparrow

SOCIABLE, constantly chattering and a little bit scruffy-looking, House Sparrows are our original feathered friends, always found close to humans. It's thought the relationship began thousands of years ago when our Neolithic ancestors starting farming grain in the Fertile Crescent of the Middle East, with these plucky little birds taking advantage of the new food sources. As humans spread across the world, House Sparrows followed. In modern-day UK, House Sparrows still nest close to humans, under the eaves of houses, in holes in buildings or nearby in hedges and trees. But it's a precarious existence. House Sparrow populations are much smaller than they once were, placing our cheery companions on the Red List of Birds of Conservation Concern.

TOP ID TIPS

House Sparrows are sturdy little birds with stubby bills. They have grey or greyish-brown bellies and brown wings with noticeable black streaks. You can identify males by their black bibs and the reddish-brown patches on the sides of their heads. Females and juveniles are lighter in tone, with a pale stripe from their eyes to the back of their heads.

Often, you will hear House Sparrows before you see them, as they swoop into the garden in noisy little groups. These talkative birds constantly chatter to each other in short, rich cheeps.

WHEN AND WHERE TO SEE

All year round in urban and rural areas.

↓ A male (left) and female House Sparrow.

Dunnock

DUNNOCKS ARE SMALL, brown birds that can easily be mistaken for female House Sparrows. It also doesn't help that until fairly recently they were quite commonly known as Hedge Sparrows. However, they are very different birds with very different behaviours. While House Sparrows are noisy, bold little birds, Dunnocks are shy and nervy. Birders commonly describe them as 'skulking'. They hop around the undergrowth, darting into bushes at the slightest disturbance. They're also more likely to be seen alone or in pairs, a far cry from the gregarious gangs that you'll often see when it comes to House Sparrows. Look out for Dunnocks hopping beneath hedges or in garden borders – but come spring, keep an eye out for a now uninhibited male singing brightly.

TOP ID TIPS

The name Dunnock is an old word, simply meaning 'brown bird', with 'dunn' an old English word for 'brownish dark grey', and it's easy to see how they got their name. They have streaky brown backs and wings, with grey heads and breasts. Males and females are very similar, with both being easy to confuse with a

female House Sparrow. Juveniles are more spotty, similar to a juvenile Robin but greyer. The bill can be a good giveaway, with the Dunnock's neat and pointy and a House Sparrow's stubby and fat.

A Dunnock is at its most distinctive when singing, when the male will take to the top of a bush and belt out a loud, high-pitched, repetitive jangle of notes.

WHEN AND WHERE TO SEE

All year round in gardens, woodlands and parks.

↑ A male (left) and female Blackbird.

Blackbird

THE BLACKBIRD is one of our best-loved and most familiar garden birds and is famed for its rich, fluty song. You may hear a male singing from a chimney pot, giving his all to a sweet, warbling melody. It's a bright, cheerful song that tells you everything is alright with the world. But disturb or upset one and it's a different story. Then they will shout loudly, a high-pitched *chink* that they repeat and repeat, often calling as they fly to a safer place.

Listen at dusk, when other birds are quieting down, and more often than not you'll hear the cross *chink*ing of Blackbirds before they settle down for the night. Blackbirds prefer to forage on the ground and are often found digging around in the undergrowth, turning over leaf litter or probing the lawn for worms.

TOP ID TIPS

The male is exactly as the name suggests, with all-black feathers. He is a decidedly smart bird with an upright stance, yellow bill and matching yellow ring around his eyes. The female is a softer-looking bird with dark brown instead of black feathers and a duller yellow bill. Females can be easily confused with the similar-looking juveniles, but if the bill is yellow, it's a female. Juvenile Blackbirds are a warmer shade of brown with fine paler spots, and also have dark bills.

Blackbirds can often be heard singing from high on a rooftop or on top of a tree, cheerfully delivering melodious phrases with a twiddly flourish at the end.

WHEN AND WHERE TO SEE

All year round in gardens, woodlands and parks.

Song Thrush

A LITTLE SMALLER than a Blackbird but no less tuneful is the aptly named Song Thrush. This super singer can be heard in late winter through to July and is often heard singing first thing in the morning and at dusk.

These speckled birds are common garden visitors, with a neat trick when it comes to eating snails. They will smash the shells of snails on a hard surface, such as a rock or a path, using it like an anvil to get into the meaty flesh inside. A collection of broken snail shells on a flat surface is a sure sign that a Song Thrush has been enjoying a feast.

TOP ID TIPS

Male and female Song Thrushes look alike with a plain brown back, speckly front and brown bill. They can be confused with juvenile or even female Blackbirds who can also look mottled on their breasts and bellies. But a Song Thrush's underside is far lighter, being creamy (whiter on the belly) with dark brown, teardrop- or arrow-shaped spots.

A Song Thrush's song sounds a little like a Blackbird's and varies from bird to bird. Try listening out for a repeated series of distinct short phrases. Song Thrushes will repeat each sequence once or more before moving on to the next one.

WHEN AND WHERE TO SEE

All year round in gardens, woodlands and parks.

Mistle Thrush

MISTLE THRUSHES can be described as the large shy cousins of the Song Thrush, more at home in open woodlands and parks than gardens. Like the Song Thrush, the Mistle Thrush is a prolific singer, with a reputation for singing as storms approach. Birds, like humans, tend to head for cover as the clouds darken and the winds whip up, but this doesn't faze a Mistle Thrush. Instead, these tough cookies seem to delight in the bad weather, singing in an act of defiance that has earned them the nickname 'Stormcock'. As with Blackbirds and Song Thrushes, Mistle Thrushes often sing from a high vantage point, so look out for them on the tops of trees or chimneys. The song of a Mistle Thrush is one of the earliest markers that spring is on its way and they can be heard singing from late January.

TOP ID TIPS

It's easy to muddle a Song Thrush with a Mistle Thrush but the spots on the belly can give it away. If the spots are rounded, then it's a Mistle Thrush, but if they are pointed then it's a Song Thrush. A Mistle Thrush is also paler on its wings and back than the richer brown of the Song Thrush, and its wing feathers have prominent white edges. Seen on the ground, Mistle Thrushes stand tall and upright, with speckly pot bellies.

The song is very like a Blackbird's but listen closely and a Mistle Thrush will stop frequently, as though pausing to think about what to sing next.

WHEN AND WHERE TO SEE

All year round in gardens, woodlands and parks.

March's challenge bird

Tree Sparrow

HEAD TO THE COUNTRY and look for the House Sparrow's rural relative, the Tree Sparrow. Like House Sparrows, Tree Sparrows are sociable birds and often seen in small flocks. Keep an eye out for them in hedgerows as you go on a countryside walk, particularly when you are close to a farm. Tree Sparrows, like House Sparrows, also tend to be found close to humans.

Sadly, Tree Sparrows are on the Red List of Birds of Conservation Concern as a result of a dramatic decline in numbers. The British Trust for Ornithology suggests that for every Tree Sparrow today, there were around 20 in the 1970s. The reasons behind the decline are uncertain. One theory is that changes in farming practices and an increased use of pesticides have resulted in a lack of food and suitable habitat.

TOP ID TIPS

Tree Sparrows are dainty-featured birds, smaller and neater than their urban counterparts. Compared to a House Sparrow, a Tree Sparrow's main distinguishing feature is its almost complete white collar around its neck. It also has a chestnut-brown crown, black bib, and a neat black spot on each white cheek. Sexes are alike.

Just like House Sparrows, these gregarious little birds chatter and chirrup as they go about their day, and it is often their calls that alert you to their presence. If it sounds like a high-pitched House Sparrow in the country, chances are it's a Tree Sparrow!

WHEN AND WHERE TO SEE

All year round in rural areas, including open country and farmland.

Monthly musings

Pigeon power

Unlike most other birds, Collared Doves can nest all year round, as long as the weather is mild enough. Typically they will make between three and six attempts a year, and like others in the pigeon and dove family, they lay two eggs each time (though it is a myth that the young are always a male and female 'pigeon pair'). Once the chicks have fledged, they can start to nest again, with the adults feeding the fledglings when they take a break from incubating the new eggs. Woodpigeons also have a long breeding period, stretching from March until October. But while they similarly lay two eggs each time, they will breed only twice a year at most.

↓ A Long-tailed Tit can use thousands of feathers to make a nest.

↗ Woodpigeon eggs. Eggs can be left unattended for short periods.

Feathering the nest

Many birds use feathers to insulate and soften the linings of their nests. Long-tailed Tits are well known for the vast numbers of feathers they use, with one Victorian ornithologist, William MacGillivray, counting 2,379 feathers in a single nest. But where do birds find all these feathers? Some such as Blue and Great Tits pluck a 'brood patch' on their bellies, which both provides feathers and uncovers a warm area of skin to incubate their eggs. Others such as Long-Tailed Tits, however, need more feathers than their own bodies could provide, so they also rely on the feathers they can find and will even pluck the bodies of dead birds. The remains of a Woodpigeon killed by a Sparrowhawk is a real treasure trove for a Long-tailed Tit in search of fluffy feathers!

Top two must-dos

1 Visit a rookery

Rooks are early nesters, building great stick nests high up in the trees in large colonies, known as rookeries. These are noisy hives of activity, as the birds busy themselves with the fraught tasks of nesting and raising chicks. Watching a rookery in March is a wonderful window into how birds annually set to work to raise the next generation.

Although they can get going from as early as January, nesting usually takes place in February and March. The bare winter branches make rookeries easy to see and they can be found anywhere from a stand of trees in the countryside to a clump of trees on the edge of a motorway. Rooks lay three or four eggs, which are incubated by the female for up to 16 days. After hatching, chicks will spend just over a month in the nest until they fledge.

Adult Rooks carry food for their chicks in their throats, in a special expandable area called a 'gular pouch', which gives their head a strange shape when it is full. Look for Rook chicks peeking out of the nest, a black splodge shouting, bill raised as the parent comes in with food. Watch too for adult-sized youngsters, all messy feathers and shaky outstretched wings, calling for food and pursuing a parent along a branch.

How to identify a Rook

Rooks are corvids, members of the crow family that also includes Jackdaws,

→ Rooks can be identified by the 'grey patch' on their bills.

↓ A rookery can be very noisy.

Ravens and Magpies. They are a similar size to the familiar Carrion Crow and similarly completely black. However, they can be told apart by their bills. While a Carrion Crow has an all-black bill, a Rook's is grey and bare of feathers at the base of the bill.

Rooks are also the most sociable of all the crow family, as their fondness for nesting in great groups demonstrates. They are often seen feeding together, typically on farmland. Here, on the ground, they look 'baggy', with an ample belly and shaggy 'pantaloons' that droop over their legs. Carrion Crows, on the other hand, are a whole lot tidier.

At the rookery it can be hard to make out the subtleties of a Rook's call, as they all seem to clamour and call with no regard for what their mate, chick or neighbour may have said. But it's worth persevering. On first hearing, a Rook's call may be dismissed as similar to a Carrion Crow's, yet there is a distinct difference and tone. A Rook gives a drawn out *caaw* or *kaah*. A Carrion Crow, though, adds a rolling 'r' and calls *krrah, krrah, krrah,* often in a seemingly angrier manner.

2 Celebrate the first Swallow

Migration has got to be one of the greatest wonders of the natural world, with the Swallow a particularly impressive long-distance migrant. Each spring, Swallows make an epic journey of around 10,000km from South Africa to the UK. It's an arduous journey across tough terrain including open sea and the Sahara Desert.

Swallows migrate during daylight hours, flying around 320km a day and feeding on insects along the way. Yet starvation is still a very real risk, as is bad weather and even being shot, making every returning Swallow worthy of wonder.

Swallows start to arrive from late March with the birds arriving in bulk in April. The arrival of the first Swallow is a hotly anticipated event for many nature-lovers and the timing of the bird's return often an indicator of annual variations in weather.

Here in the UK, the first Swallow also signifies the start of summer and, although we're warned that 'one swallow doesn't make a summer' (see page 57), a sighting assures us that warmer days are on the way. Each Swallow aims to return to its birthplace to nest each year, often stopping to feed and drink at wetland sites on the way there. So visit your nearest waterway or the place where you last saw a Swallow and celebrate the winged wanderers' return.

↑ A Swallow has distinctive tail streamers.

But watch out for those that never went away

In 2022, the British Trust for Ornithology (BTO) reported that some Swallows had opted to stay in Britain for winter rather than migrate back to their wintering grounds in South Africa. Swallows migrate back because the much colder British winter makes it too cold for the flying insects that they feed on. However, a run of milder winters brought on by climate change is thought to have led to a small number of birds attempting to spend winter in the UK. Migration takes vast resources of energy and involves huge risks, so it easy to see why some might hedge their bets.

In 2022, data from the BTO's BirdTrack survey of birdwatchers' observations found almost 100 reports of up to 12 Swallows seen between 1st January and 1st February. As might be expected, most were sighted in the warmer south and south-west of Britain, and also in Ireland. The BTO suggests that overwintering Swallows is something that we may see more frequently in future years.

How to help

Feed the birds

It's not just in the depths of winter that garden birds could do with a little extra help. During nesting time and later when the chicks hatch, your busy garden birds may well thank you for some supplementary food. Birds rely on there being an abundance of particular types of foods when their chicks hatch. Tits and Chaffinches, for example, time their breeding to match an increase in caterpillars. Blackbirds and Song Thrushes, on the other hand, are hoping for plenty of earthworms. Unseasonal weather, whether a severe cold snap, a very high rainfall or long, hot days can all lead to food shortages. Cold, wet weather can mean a lack of insects, while long dry spells can make the soil too hard to dig for worms.

You can help by putting out food such as bird seed mixes. Chopped soft apples, pears, bananas and grapes are also good. But avoid giving peanuts, fat and bread since these can be harmful to chicks. It's also a good idea to soak dried mealworms, as these are easier for chicks to swallow.

Crucially, however, be sure to clean your feeders, bird tables and bird baths regularly (see page 7). Garden birds can be susceptible to disease, but keeping places where they congregate clean can help keep them healthy.

↓ Coal Tits will eat a range of seeds, fruits and nuts.

Record your sightings

BirdTrack is a citizen science project that enables you to record and share your bird sightings in Britain and Ireland. You can contribute all year round. All you need to do is make a note of the birds you see, whether out birdwatching or at home, and enter what you saw using the BirdTrack app or via the British Trust for Ornithology (BTO) website. As well as adding your own records, you can take a look at what others are seeing and view trends for where and when different species have been seen. It also acts as your own personal record, storing your sightings.

BirdTrack provides valuable data and is helping to map when our migratory summer visitors such as Swallows arrive and depart. This could prove particularly useful when looking at how climate change is impacting on bird behaviour.

Similarly, it is helping to map the habits of winter migrants, birds such as Redwings and Fieldfares. It also provides information on less common birds, such as Crested Tits and Nightingales, which are found in only a few areas in the UK.

BirdTrack began in 2002 as Migration Watch and was originally set up as a tool to help map bird migration. But in 2004 it was renamed BirdTrack and expanded to cover all seasons. In 2021, BirdTrack was extended further so that people could record other animals. You can now add your sightings of mammals, amphibians, butterflies, dragonflies and reptiles.

BirdTrack is organised by the BTO in partnership with the RSPB, Birdwatch Ireland, the Scottish Ornithologists' Club and the Welsh Ornithological Society.

Myth of the month

Wishing for Swallows

The saying 'one swallow doesn't make a summer' suggests that just because one good thing has happened, life is not about to get a whole lot better.

It's a theme described in the fable 'The young man and the Swallow' attributed to the Greek slave Aesop who lived around 2,600 years ago. A young man, down to only a cloak after losing all his other belongings through gambling, sees a Swallow. Believing that this means spring and warm weather is round the corner, he decides to join another game and puts forward his cloak as a wager, thinking he will no longer need it. But he loses and a snowstorm blows up. He sees the Swallow now dead from cold and, freezing himself, regrets his rash decision. This became the Greek proverb 'one Swallow does not make a spring' described by Aristotle. The interchangeability of spring and summer is easy to understand when you consider the relative temperature of summer in northern Europe and spring in the Mediterranean.

It's a neat phrase and rooted in a natural truth. Swallows and other summer migrants don't all arrive together and there are always a few outliers who arrive ahead of the crowd. These early birds may benefit from securing a better territory than the latecomers, but may also face a last spell of wintery weather. Although we would be wise to heed the proverb too, we can be sure that given a bit more time, the rest of the Swallows will eventually turn up.

↓ Wisdom hidden in a Swallow's ways?

APRIL | 4

Songs of love and war

April is peak birdsong month. Go out anywhere and your ears will be enveloped by delightful sounds, especially at dawn. But for the birds themselves, the message of song is not a gentle one; it's to declare ownership of a territory, as well as to command the attention of the opposite sex.

IT'S NOISY OUT THERE. The dawn chorus is in full swing, and in woodlands and scrubland not a minute goes by when the air isn't filled by a bird with something to say. Even the most nature-unaware people cannot help but notice the rising choir, even if they resent being woken up!

Songs of war

We shouldn't be fooled that it's harmonious, though. Birdsong isn't an expression of joy, it's an expression of challenge. Almost everything you are hearing is an unambiguous proclamation of territorial rights. Throughout the year, birds make a variety of sounds, often short one-note or two-note expressions of alarm, contact and so on – these are bird 'calls'. Song is quite different. Most songs could be described as sentences, or even paragraphs, as they are more complex and variable than calls. These songs have biological purpose.

When a bird wants to breed, it must find a territory, one that can encompass a nesting attempt. This is almost always the male's primary role, not the female's.

→ The song of the Blackbird is glorious, and individuals improve their repertoire year on year.

← Willow Warblers sing energetically the moment they arrive in Britain in spring.

A male's first task is to sing, making a statement of ownership. As the breeding season progresses, birds fight, in song, over territory. There are winners and losers, and the losers may not breed at all if they cannot get their song together.

Songs of love

The females make a keen audience for this competition because, as the males sing, they can listen and assess the quality available. Although it isn't always obvious to us, songs differ greatly between males of the same species. Some sing more vigorously, some sing for longer periods and others sing a more interesting song, with a better repertoire. So long as a song is recognisably of the right species, there is room for variety. A male's song really is akin to a dating profile, but he cannot fake anything; his experience, health and inventiveness are exposed for all to assess. His singing is a crucial tool not just for acquiring that territory, but a mate, too. No wonder birds sing with such vigour!

For us, it is all delightful. Despite the fact that the birds themselves are stressed, birdsong has the opposite effect on us, actually calming us down. A study by King's College, London, found that even in a studio, people subjected to a few minutes of a

particularly euphonious song, such as a Blackbird's, exhibited lower stress levels. This is a gift for busy people.

If you have a garden, or easy access to any open space, it's time to revel. You will probably hear the wistful voice of a Robin, every phrase different, or the simple repeated warble of a Dunnock. Wrens may shout at length from their tangle of vegetation close to the ground. Song Thrushes sing at all hours, stridently repeating every phrase a few times before moving on to the next. Blackbirds give a soothing, fluty harmony. Great Tits chime their repeated *teacher, teacher, teacher*, while Blue Tits make a silvery mutter. Chaffinches produce a cheerful chatter, while Greenfinches utter canary-like trills interspersed with

→ Many birds (such as this Greenfinch) select high perches to help their song to carry.

↑ You can hear the Sedge Warbler singing from early April onwards.

drawn-out wheezes. The local Great Spotted Woodpecker makes its own singular 'song' by drumming – pecking loud and fast against sonorous wood (see page 16). The garden, with its limited species, is a great place to learn songs.

Dawn chorus

We all know that birds sing lustily in the darkness before dawn, but you might be surprised to know that, for scientists, this has long been a puzzle. We know why they sing, but why this cacophony before the day begins? Studies suggest that in most situations, it's simply a roll call. Territory-holders might die overnight or wake up feeling too ill to sing. In such situations, competitors quickly notice the change, and could take over the territory. A female whose mate is AWOL could take the chance to copulate with another bird. If so, it will happen at dawn. Females typically mate and lay an egg first thing in the morning if they are in mid-clutch.

It happens that dawn is a great time to transmit your song, and the darkness prevents other activities such as feeding (insects won't be warm enough to move). Predators won't be active, either. So, it's a good time to make your mark on the day.

Incomers and outgoers

April isn't just a noisy month; it is also the peak arrival time for most of our summer visitors. If there wasn't enough joy already for an indulged birdwatcher, now almost every day something new is arriving to join spring's potent mix. This is the month when you have the chance of hearing your first Cuckoo, of seeing your first Swallow if the last days of March did not bring one, and of being challenged by warblers of all kinds. There is a warbler for every week: a Blackcap in the first, Sedge Warbler in

the second, Reed Warbler in the third and Garden Warbler in the fourth, although this isn't cut and dried. But April really is like a party warming up; there is a sudden burgeoning of arrivals and new personalities.

Amidst all the excitement, it's very easy to forget that there's a northward movement in April away from us, too. Many of our favourite winter birds, such as Redwings, Fieldfares and Bramblings, not to mention a whole host of ducks, geese, gulls and waders, quietly leave. We are usually too bowled over with spring's excesses to notice.

Egg month

Singing is the loud April revolution, but egg-laying is the quiet April revolution. It has to be. Suddenly, attracting attention is the last thing the birds need. A nest is, out of necessity, carefully hidden

↑ Great Tits lay their only clutch of eggs in April.

away because the clutch and the female are now at one of the most vulnerable points in their lives. You shouldn't notice anything, but there are clues. You might spot male Robins feeding their mates to help out, or you might spot a female bird eating snail shells to help her form her eggs. But the emphasis is now on privacy.

It's the time of year when most of our inland resident birds lay their first, and sometimes their only, clutches, at least away from northern Scotland where things start later. All the British tits lay their clutches in April, and most other familiar birds from the garden, for example, have things well underway. Even some summer visitors manage to arrive, pair up and lay a clutch before the month's end, which is pretty impressive. The spring is progressing fast.

Birds of the month

Chiffchaff

IF YOU SPEND any time in woodland in spring, you will certainly have heard the Chiffchaff, even if you weren't aware of it. Its seemingly endless *chiff-chaff-chiff-chaff...* song is part of the soundtrack of the season, especially in taller deciduous woodlands. The song is so monotonous that you wonder if the male ever gets tired of singing it! This is a small bird that is mainly found in the woodland canopy, restless and constantly moving about, fluttering from branch to branch. It has a habit of using dead snags, or boughs still without leaves, for song posts. Curiously, it has a fidgety habit of constantly wagging its tail downwards.

TOP ID TIPS

It is Blue Tit-sized and equally active, but you'll soon see that it really lacks any markings except for a pale stripe over the eye. Essentially it is a greenish-yellow colour, darker above and paler below, with males, females and juveniles all alike, though by late summer, after the breeding season, adults can look very scruffy. Look out for the white eye-ring, more obvious than that of the very similar Willow Warbler, which has paler cheeks. It also has dark legs (flesh coloured in Willow Warbler) and shorter wings than its pointed-winged relative.

WHEN AND WHERE TO SEE

It is mainly a summer visitor, arriving in March and departing by October, but some are present in winter, too, probably migrants from northern Europe. It is very common in deciduous woodlands.

↓ Chiffchaffs often sing from leafless branches.

Blackcap

THIS MONTH, Blackcaps flood into the UK and their songs soon accompany almost any woodland walk. They have a clear, sweet whistling phrase that starts hesitantly and then improves, but it can be difficult to pick out from the rest of the bird chorus. Blackcaps are restless birds, but are hard to see well because they hide away in the low and middle branches of trees. Besides woodland, they are common in tall scrub, parks and large gardens. Many people know them better from their visits to bird tables in the winter for suet and fruit, where they are noticeably aggressive towards other birds. But they are far more common and widespread now than during the winter. One of their interesting quirks is that the male makes several nests, which the female inspects and often rejects completely, deciding to make a new one!

TOP ID TIPS

At first sight a small brown bird, but it is noticeably greyish, especially around the head and neck. The male has a neat black skullcap, but the female's is chocolate brown, as is the juvenile's.

WHEN AND WHERE TO SEE

Mainly a woodland bird, present from April until early October; in autumn it is easy to see feeding on elderberries and other fruit. A small number (migrants, mainly from central Europe) winter here.

↙ The Blackcap has a distinctly grey tint to its plumage. This is a male. The female Blackcap (inset) has a brown cap.

Woodpigeon

IF YOU CARE to watch, Woodpigeons provide terrific entertainment. At this time of the year, and right through the whole summer, they are always showing off. Take their display-flight. It involves a bird flying up at an angle, flapping its wings and then quite suddenly stalling, until it opens its wings, spreads its tail and glides down in the same direction to a new perch. There are days when they never stop doing this. They also have a deliciously fulsome song of deep, theatrical coos in series of five – *My TOE hurts, Betty!* – which they often sing from rooftops and chimneys. If not displaying like this, members of a pair sit together and preen each other in shameless Public Displays of Affection. And, as you may remember, their nests are embarrassingly awful (see page 43). These birds are characters!

TOP ID TIPS

A big pigeon with an obvious white patch on the side of the neck. In flight, there is a big white stripe across the middle of each wing. The breast has a pleasing flush of pink and the neck an iridescent green patch. The bill is pink and the eye whitish. Beware: young birds lack the white patch on the neck! They also have darker eyes.

WHEN AND WHERE TO SEE

It's everywhere, all year round.

Collared Dove

IT SEEMS THAT suburbia and Collared Doves were made for each other. You see more Collared Doves in built-up areas than anywhere else, where they flourish in gardens and parks. Farms can also be attractive to them. The signature sighting is of a bird on an aerial or rooftop, singing its dirge-like, three-note coo, sometimes rendered *Un-I-ted*. They are common visitors to bird tables, where you can appreciate their delicate creamy colour and slim, long-tailed appearance. Their display-flight is eye-catching, involving a steep climb upwards, a clap of the wings and then a spiral down on spread wings and tail. They often nest inside garden shrubs and can be remarkably productive, raising up to six broods of two chicks a year.

TOP ID TIPS

The most obvious feature is a thin black half-collar around the neck, fringed with white, although this is missing on juveniles. It is pale fawn-coloured above and creamy below, with a white-tipped tail. When flying, it has a definite 'flicking' action to the wingbeats, quite different to the full wingbeats of a pigeon. Besides the song, it has a curious trumpeting croon of a call, like a party-blower.

WHEN AND WHERE TO SEE

Widespread and common all year round throughout the lowlands of the UK, usually around towns and villages.

Wren

WOULD YOU BELIEVE that the Wren is the UK's most numerous wild bird? Ten million pairs pack into every corner of the country. But you shouldn't be embarrassed if you've never seen one, because you'll be in good company. They are easy to overlook, not just because they are tiny (although not quite our smallest bird – see Goldcrest, page 14), but also because they spend almost their entire lives hidden in the secret passageways of low vegetation and dense shrubbery. They only emerge to dash low across a path on a whirr of wings or, more often, to sing their shouty song from a bush-top perch. When singing, the bill is open wide and the whole body trembles with the effort. At this time of the year you can hardly go anywhere without this song accompanying you.

TOP ID TIPS

A tiny ball of feathers with a sticky-up tail. A closer look reveals barred plumage and a pale eyebrow. It's mouse-like, but a mouse that roars: the song is long, loud and trilling, and heard almost all year, always low down. It also makes a call, an irritable *teck*.

WHEN AND WHERE TO SEE

Very common everywhere all year round, from gardens to the moorland of isolated islands.

April's challenge bird

Feral Pigeon

OK, IT'S HARDLY a challenge to see the humble pigeon, is it? But there are times when it's a good idea to look at a familiar neighbour in a new way – and even to appreciate it a little. It has, for example, a showy courtship dance, and mated pairs show tender bonding behaviour. They are also devoted parents, but these family units exist within a wider social group with its own complex hierarchy. The pigeon has many stories to tell.

But what is it? Most people refer to 'Feral Pigeons', which are descendants of a wild bird known as a Rock Dove or Rock Pigeon, which still lives on cliffs on remote northerly and westerly parts of the British Isles. Rock Doves have been domesticated since at least Neolithic times, for food and for carrying messages, and Feral Pigeons are simply free-flying versions of these, nesting on tall buildings as a substitute for their ancestral sea-cliff homes. Centuries of crossbreeding have produced many forms, and most flocks of Feral Pigeons are easily told from other pigeons by their multiple colours and patterns, each individual distinct from the rest.

TOP ID TIPS

Although their plumage colour and pattern varies greatly, with some individuals looking nothing like our other wild pigeons, plain grey Feral Pigeons can look quite similar to Woodpigeons and Stock Doves. They differ in having orange eyes (browner in juveniles), a black bill and striking white underwings.

WHEN AND WHERE TO SEE

Towns and cities, and also on wilder sea cliffs, all year round.

Monthly musings

← Blue Tit nestlings get a better start in life by having aromatic plants placed among the nest material.

Blue Tits disinfect their nests

Nests are messy, unhygienic places, especially Blue Tit nests, often home to 10 or more chicks all closely packed together. But Blue Tits have a way of disinfecting them and purposely bring in aromatic herbs, including lavender, mint, Red Deadnettle and Ground-ivy, to line the structure. Individual female Blue Tits, who build the nest, have their own preferences. Their relatives Great Tits and Coal Tits don't do so. Research has shown that the herbs help to reduce the bacteria in the chicks' bodies, making them grow faster and have more red blood cells, a good predictor of future survival.

Different bills for different foods

Did you know that you can tell a bird's diet from its bill shape? It makes sense, since the bill is the tool that most birds use for obtaining as well as consuming food. Small birds tend to be primarily seed-eaters or insect-eaters. Seed-eaters have broad, conical bills, ideal for crushing seeds internally, while insectivorous birds have thin bills, perfect for picking and probing in vegetation. Predatory birds have sharp-edged, hooked bills with a deadly tip that can help keep meat in place. Fish-eating birds tend to have long, straight bills to reduce water resistance, sometimes with hooks or saw-like serrations to help them grip a slippery meal. Very long, slender bills are often used for probing into mud.

↓ Seed-eaters such as the House Sparrow (left) have thick, conical bills. The Dunnock (centre) has a typical thin, fine insect-eater's bill, ideal for picking insects off vegetation. Birds of prey such as Red Kites (right) have hooked bills.

Top two must-dos

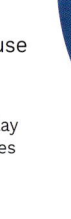

1 Watch a Buzzard display

If raptors had personalities, the Buzzard would be an extrovert. While many birds of prey are secretive and rarely seen, the Buzzard is always in the limelight. It is showy, regularly soaring high in the sky and making its wondrous, atmospheric call-note. Neighbouring birds often meet and soar together; they do this at their territorial boundaries, because

→ The courtship display of the Buzzard includes spectacular mid-air plummeting dives.

the Buzzard is also by far the most territorial of our raptors. Skirmishes occur frequently.

At this time of the year – indeed from February onwards – the Buzzard adds an extra layer of theatricality to its repertoire by performing jazzy aerial displays, intended mainly for the eyes of their mate. These antics can go on for many minutes and are well worth watching. You will also be able to enjoy the accompaniment of the Buzzard's mew, which will be frequent. It has a wild ring to it that is hard to describe, but easy to appreciate.

The male Buzzard's display involves flying up to a steep height, which can be 300m or more, and then dropping down, its wings partially held in. It will sometimes twist round on a vertical axis on its downward plummet, but once it has reached roughly treetop height it will suddenly pull out of the dive and use its considerable momentum to swoop up again. This swoop will give it lift and, once it has reached a certain height, it can then loop down again and repeat the whole thing. Sometimes it will give 10 or so dives and upward swoops in succession. After a short break, perhaps perched, it will be ready to go again.

Sky-dancing and ground feeding

Often, the female will join her mate in the air, in which case any displays will take place after much mutual circling and calling. The male weighs less than the female, so he can more readily lift up quickly well above her, and will then aim his dive down towards her, making sure he only just misses as he plummets past. He will also usually call when they are close to each other, as if to say: 'Look at me!' The effect of these large birds dancing in the sky is a truly magical one.

Buzzards usually perform only in good weather, and damp rainy days cramp their style. But in horrible conditions you have the chance to see the Buzzard in a very different light. On such days, these predators often spend large amounts of time walking over the mud in fields, trying to catch worms. It's nutritious, but a bit of a come down.

→ Grey Herons build large stick nests in the treetops.

2 Enjoy the antics of a heronry

The Grey Heron is one of the year's earliest nesting birds. Pairs often attend their treetop colonies from late January, and they have been recorded laying eggs as early as 19th February. On average, though, the clutch is completed by mid-March and, after almost a month of incubation, that means that the youngsters will be hatching out about now. Heronries will be busy places.

It always seems incongruous that a large, long-legged bird that spends much of its time paddling in shallow water should somehow nest in the tops of trees. It must be awkward for the legs and feet to move among the branches and place sticks to make the very large platforms. They must also collect this material from the ground or break sticks off trees; many take the shortcut and simply steal material from neighbours. Anyhow, the nest is at least safe, so long as the chicks don't fall down from nests that can be 25–40m above the ground, which they sometimes do.

Herons lay 4-6 eggs and, once they all hatch, the youngsters are brooded for about three weeks. During this time, one adult needs to go to find food for the growing family. Feeding visits are hilarious. Every time an adult approaches, the chicks respond as if they have never been fed before. They violently grip the parent's head and neck in their bill and it looks for all the world as though they are stabbing them. All chicks compete, but because not all eggs hatch at the same time, the brood consists of different-aged chicks, and younger birds can be outcompeted. They often die as a result.

Grey Herons occasionally nest singly, but far more often in company with others. The colonies vary greatly in size, but there aren't usually more than about 10 nests, at an average of three per tree.

If you do visit a heronry, one thing you won't forget is the noise. The birds make a variety of sharp quacks and grunts, along with the more familiar, testy *krank* call. The noise intensifies when an adult flies in to feed young, who respond with excitement. The neighbours might also make some complaining calls when the visitor disturbs their peace, and there is an enjoyable kerfuffle.

↓ Herons have large clutches for such big birds, often 4-6 eggs. Incubation begins with the first egg and eggs are laid at intervals, so young are of different ages. They compete vigorously and the youngest chicks rarely survive.

↘ Although not as celebrated as bees, hoverflies are fantastic pollinators and are a sign of a healthy garden.

How to help

Garden for wildlife

It's easy to feel discouraged about the state of bird and other wildlife populations in the UK. But while the stated declines are only too real, many of us are unaware how easy it is to help nature and to make a difference ourselves. If we are fortunate enough to have a garden, or an open space over which we have some sway, we can help turn the tide. Private household gardens in the UK amount to an area the size of the county of Suffolk, and the influence of wildlife-friendly gardening stretches to every corner. When wildlife is a priority in a garden, the effect is dramatic.

One of the ironic things about wildlife gardening is that you can do a lot of good by doing nothing at all. It's now recognised that messy corners are a good thing, indeed essential.

For generations, the neatest and tidiest gardens have been the most revered, right down to viciously pruned hedges with stark straight edges, lawns suitable for a bowling green, and flowerbeds purged completely of weeds. People still think this way is 'best', but it is no more than deep-rooted fashion, and is just a way of bringing the indoors outside. Most people like tidy houses, but the garden is part of the outside and should be different. If you take a look at any thriving ecosystem, it has haphazard lines and unruliness; that is its charm and essence. Garden wildlife usually benefits when the stranglehold of tidiness is broken.

So, while everybody has different tastes that need to be respected, there is no doubt that the garden's minibeasts

– and, logically, the birds that eat them – thrive when you allow wildflowers such as Daisy, Dandelion and Self-heal to grow on the lawn, leave dead wood (immensely beloved by beetles and many other invertebrates) in situ (if not dangerous) and allow genuinely helpful plants such as Ivy to run as rampant as possible.

Please do swerve well away from the harmful plastic lawn craze; don't even contemplate it. And if you can avoid using any pesticides or herbicides, then you would receive a great deal of affirmation if your birds, invertebrates, mammals and everything else could talk to you.

The wonderful thing is that gardening for wildlife isn't just your delight for today, but your legacy for tomorrow. Today you could plant a berry-bearing bush, to feed birds in autumn for generations.

CREATE A WILDLIFE GARDEN

The delight about wildlife gardening is that, alongside the things that you don't do, there are many things you can do.

- Plant flowers that the pollinators revel in, and make sure that there's a year-round plenty, from crocuses to Michaelmas daisies.
- If all you have is a windowsill, put out plants such as thyme and lavender.
- Plant umbellifers, such as hogweeds, as a dancefloor for randy insects.
- Provide every kind of home you can (see page 163).
- Leave seedheads alone so insects can use them.
- Install a compost heap.
- Dig a safe pond.
- Plant a fruit tree.
- Make a place for a meadow, however tiny.

↑ Dandelions are not appreciated by most people, but they have abundant supplies of nectar.

Myth of the month

The Cuckoo predicts your fate

If you are lucky this month, you might hear your first Cuckoo of the year (see also page 89). However, if you do have an encounter with one, beware, because the Cuckoo has news for you.

For centuries the well-timed appearance of this loud, evocative songster, the twin herald of spring along with the Swallow, has given rise to sayings about how the year will turn out. One of the starker legends of the Cuckoo as fortune-teller is that the very first time you hear the singer in spring, the number of repeated *cuckoo*s reveals the number of years you have left of life, as in the saying: 'Cuckoo, my little Cuckoo, how many are my years?' More positively, perhaps, another saying suggests that the number of *cuckoo* repetitions is a direct prediction of the year's forthcoming income. That should help you prepare well. Just don't tell the tax office.

It also matters, apparently, how you experience the first Cuckoo. If your stomach is full when the voice of the unseen singer touches your ears, the year will go well. If you have a coin in your pocket (OK, this might need to be modified to a debit card), then that's excellent news, and you will enjoy prosperity.

There is quite an irony to all these stories because there is one thing that the arrival of the Cuckoo most certainly guarantees: bad fortune for small insectivorous birds.

↓ The Cuckoo arrives back in the UK in April.

MAY | 5

Long-distance flyers

Migration is one of the wonders of the natural world, so this month take a moment to marvel at the journeys migrant birds make and enjoy the aerial acrobatics of two much-loved visitors to the UK, Swifts and Swallows.

MAY IS A MONTH to look up in awe as by now the summer migrants will have arrived. If you're in an urban area, look to the skies for shrieking Swifts, zooming above the rooftops or swirling scythe-like high overhead. Somewhere more rural? Then look for Swallows, their telltale tail streamers flowing behind them as they acrobatically twist and turn, snaffling up insects.

Around 4,000 bird species worldwide migrate, some two-fifths of all known species. In the UK this actually increases to around half of our bird species as the seasonal switches between hot and cold weather sees so many birds on the move. The reason behind all this movement is simple – food! It's this pursuit of food that brings a wealth of birds to the UK as spring arrives.

Flying in for food

By May, most of our summer migrants will be here. These are birds flying into the UK, choosing to spend their summer here to nest and raise their chicks. Just as our resident birds' breeding is timed to coincide with an abundance in food, migrant birds are also hoping for rich pickings and their arrival aims to capitalise on an increased availability of their favourite food.

Swallows are one of the first arrivals, with most flying in from South Africa in April. These long-distance flyers are insectivores, and happy to take a bite where they can get it, whether airborne

← Swallows typically nest in barns and are also known as Barn Swallows.

↑ Swifts can form fast-flying 'screaming parties'.

insects flying high or those closer to the ground. Swallows are particularly at home darting above water, taking emerging mayflies and other aquatic invertebrates. Next up are the House Martins, who typically arrive towards the end of April, followed by Swifts who arrive in early May. Both these birds are on the hunt for airborne invertebrates, such as spiders, aphids, flying ants, mosquitoes and other flying insects, generally feeding at a higher altitude than Swallows.

We are also joined by a whole host of other birds, from tiny but tuneful warblers, such as Garden and Willow Warblers, to mighty birds of prey, including fish-eating Ospreys.

Endurance athletes

The distance flown by our summer migrants can be truly breathtaking. At around 45g, a Swift weighs less than a slice of bread! Yet these fast-flying birds have one of the longest migrations in the world, flying around 22,000km every year from and to equatorial and southern Africa. They are endurance athletes, able to fly an average of 570km per day. Swifts are so adapted to flying that they do almost everything on the wing, including mating and sleeping, and in

2016 researchers found that adult Swifts can go a full 10 months without landing.

Super senses

Stamina aside, the science behind how birds navigate such huge distances, often returning to the same spot each year, is also awe-inspiring.

Birds are thought to use a number of visual cues to find their way, including the stars, position of the sun and landmarks such as the coastline or even a motorway. Homing pigeons on a familiar journey home have been found to follow a road, even in preference to a quicker route. It's thought that following the road takes some of the stress out of the journey. They don't need to think about where they are going and so can focus on flying fast and avoiding predators instead. Sense of smell has also been shown to be an important tool for migration among some seabirds.

But these navigation tools are nothing when it comes to a bird's extra super-sense. Birds also have the ability to sense magnetic fields and use this to help them migrate. While we can use a compass to find north, birds essentially have an in-built magnetic compass that enables them to both work out directions and figure out their location by tapping into the earth's magnetic field. Scientists are continuing to unravel exactly how this works, and researchers have discovered specialised cells in certain birds' eyes that can help them to effectively 'see' magnetic fields.

Familiarity with the route is certainly helpful, and some birds, such as geese, travel in groups or family parties, which enables youngsters to benefit from their seniors' prior experience. But what about birds migrating alone, for the first time? How can a Cuckoo, for example, brought up by a species that doesn't migrate, such as a Dunnock, hope to find her way to the right place in Africa, where she will find the resources she needs?

There is evidence that the timing of departure and the initial direction of travel are determined genetically. A single genetic mutation in the Blackcap, for example, can switch the birds' preferred orientation at migration time from south-west to east. However, there is still debate on the extent to which any birds have the innate ability to navigate all the way to their destination. Some sort of instinct does seem likely. But it's also suggested that birds can learn from others, even without travelling in close-knit groups, with young birds tapping into the calls made by adult birds as they migrate nearby.

↑ Ospreys are birds of prey that migrate from Africa.

Arriving under cover of darkness

While both Swallows and Swifts can be seen migrating by day, many smaller birds such as warblers migrate at night. One of the advantages of this is that they can spend the daylight feeding and refuelling. Flying at night also confers a number of other advantages. There is a lower risk of predation at night, with raptors such as Sparrowhawks and Peregrines generally tucked up in their roosts.

The air is also cooler, making it more comfortable to fly, and it is generally less windy, minimising the chances of being blown off course. Birds that migrate during the day tend to be stronger flyers and better equipped for dealing with the stronger thermal air currents during the day.

Navigation may also be fairly straightforward as they set off. Birds that migrate at night begin their journeys at dusk, just as the sun is setting. To travel north to the UK, they just need to keep the point of the setting sun to their left, until they can see it no more. Once the sun has set, it's likely that the birds switch to using the stars and their other support senses. Migration will likely take place over several nights, with these nocturnal flyers flying for a few hours each night before landing and settling down to sleep until dawn.

The magic of migration

There is still much we don't know about migration, which makes the annual return of summer migrants even more magical. Whatever you do this month, take a little time out to look up in wonder at our migrant birds, those most superb of flyers who have dashed over desert and soared over sea to reach us.

↑ Garden Warblers are more commonly seen in woodland than gardens.

Birds of the month

Swift

SWIFTS ARE FAST-FLYING wonders who live life on the wing, only landing to nest and raise their chicks. They migrate to the UK in spring, arriving from late April. Birds start heading back to Africa from mid-July. By September their brief but spectacular stay will be a distant memory.

Sadly, Swifts have suffered huge declines in recent years, with numbers breeding in the UK plummeting by 60 per cent between 1990 and 2020. There are a number of factors at play, including a decline in insects and a lack of suitable breeding sites once they get to the UK. We can help turn this round by putting up Swift nest boxes (see page 110), and there are some wonderful Swift groups who can help with this, such as Swift Conservation.

↑ Juvenile Swift.

← Adult Swift.

TOP ID TIPS

Look for their long, narrow wings, giving a scythe-like silhouette as they dart and dive in hot pursuit of their insect prey. In the sky Swifts look black, but they are actually a dark brown, with males and females alike. If you can catch a good look at them, you can make out the juveniles by their pale foreheads.

A Swift's call is a thin, drawn-out screech and is unmistakable.

Sometimes called devil birds, these birds can form screaming parties as groups fly around at rooftop height, shrieking loudly.

WHEN AND WHERE TO SEE

April to July in towns and villages.

Swallow

SWALLOWS ARE MORE properly known as Barn Swallows. These are birds that are found almost worldwide, and this internationally recognised name references their long-held habit of nesting in farmland barns. Swallows are the summer birds of the countryside, nesting on farms and in villages, creating mud-cup nests in open barns, under bridges and in other open shelters. Like Swifts and martins, Swallows are insectivores that feed on the wing, swooping and diving to catch their prey.

The early pioneers will arrive from their winter homes in Africa as early as March, soon to be followed by more in April. Their return brightens up the barnyard, as Swallows constantly chatter while they nest and raise their chicks.

TOP ID TIPS

These elegant summer migrants have distinctive forked tails with long streamers. They have glossy blue-black wings and back, a deep red face and are largely buff-white underneath. When seen flying high, Swallows can look black and can easily be confused with Swifts and martins. But keep an eye out for their tail streamers and you can be sure it's a Swallow. Male and female are alike except for the female's shorter tail streamers, while juveniles are a little paler with no elongated tail streamers.

Swallows call frequently with noisy little cheeps and squeaks, a little reminiscent of budgies in a pet shop. Their alarm call is similar but with even more urgency and often repeated twice, rather appropriately sounding like *Flit! Flit!*

WHEN AND WHERE TO SEE

March to September around farmland and rural villages.

House Martin

NAMED FOR THEIR tendency to nest on or near our homes, House Martins arrive in the UK in April, returning to Africa in September. Like their Swallow cousins, they build mud nests, typically on human structures, especially under the eaves of houses.

Most commonly found in towns and villages, House Martins are now on the Red List of Birds of Conservation Concern, having declined dramatically in recent years. One problem is a lack of suitable nest sites, as newer buildings no longer provide the nesting opportunities favoured by these charismatic birds. It's also thought that a decline in insects is negatively impacting on House Martins. We can help House Martins by installing special nest boxes on our homes.

TOP ID TIPS

A House Martin is a little smaller than a Swallow. Both species have dark wings and a forked tail, so can look similar in flight. One thing to look out for is a white square on the lower back, which only the House Martin has. House Martins also lack the tail streamers of Swallows. Male and female House Martins are alike, and juveniles are slightly drabber.

These are noisy, very chatty birds that call to each other nonstop. Listen out for a frequent *purr*, which sounds as though they are blowing a raspberry.

WHEN AND WHERE TO SEE

Between April and September in towns and villages.

Sand Martin

SAND MARTINS can easily be mistaken for House Martins. The clue, however, is in the name. While House Martins are more likely to be seen near our homes and buildings, Sand Martins nest in colonies in sandy or earthy banks, usually near water. Former gravel pits turned nature reserves can be particularly good spots for them. Look for Sand Martins skimming low over the water, chasing insects and giving their buzzing call.

Sand Martins can arrive as early as March, and most leave in early September. Happily, these birds have increased in recent years, possibly thanks to the creation of numerous artificial nest banks with nest holes. In 2015, being no longer considered a bird of conservation concern, Sand Martins moved from Amber to the Green List.

TOP ID TIPS

Sand Martins are smaller than House Martins and browner in colour. They have grey-brown wings and back, and are white underneath. Unlike House Martins, they have a brown collar on their breasts, and no white rump patch. Males and females look the same; juveniles are similar but with a duskier and scaly-feathered appearance.

A Sand Martin's call is a rather strange electronic squelch. Listen out for it as you watch them darting to and from their nest sites. They are especially noisy around their colonies, calling fast and frequently.

WHEN AND WHERE TO SEE

Between March and September around lakes, reservoirs, large rivers, gravel pits and the coast.

Skylark

SKYLARKS ARE FOUND in the UK all year round, but it's spring and summer when they can be most admired. Head to open country, such as a heath, meadow or especially farmland, and listen for a buzzy, fast-paced twittering song playing out in the skies above.

You will likely hear a Skylark before seeing one. But on hearing the song, look up and see if you can make out a brown bird with wings outstretched. As the male sings, he will climb higher and higher until he is pretty much impossible to see. He'll continue to sing as he descends, announcing his territory and his fitness to attract a mate and deter his rivals.

TOP ID TIPS

On the ground, Skylarks look brown, with streaky wings and back. They have a small crest on the top of their heads, which they raise when excited or alarmed. Males and females are similar, while juveniles lack crests and have a spottier plumage.

A Skylark's song is a continuous stream of varied whistling and chirruping and can last a whopping 15 minutes. To some the song sounds like the action-packed bleeps and squeaks of a 1980s computer game, while others liken it to the hyped-up electronica of an Ibiza DJ. For the more classically minded, *The Lark Ascending* by Ralph Vaughan Williams is a musical tribute to this little bird's song.

WHEN AND WHERE TO SEE

All year round in open countryside, including farmland and moorland.

May's challenge bird

Spotted Flycatcher

ONE OF OUR MOST charismatic summer visitors has got to be the Spotted Flycatcher. A little bigger than a Robin, it is rather unremarkable-looking at first glance and often put in that difficult-to-identify birdwatching category of 'little brown job'. But beneath that dull brown colouring is an artful hunter whose mesmerising skill could have you watching for hours.

Visit a woodland or park between May and August and look for a bird perched on a low branch in an upright stance, primed for action. It watches, beady eyed, then darts off suddenly, catching a fly or a butterfly with a snap of its bill, then flies back to the same perch to eat it. Like many of our summer migrants, however, this dashing assassin has greatly declined in the UK and is now on the Red List of Birds of Conservation Concern.

TOP ID TIPS

Spotted Flycatchers are subtle birds, most easily identified by their behaviour. They have grey-brown backs, wings and head, with pale grey-white bellies. The sexes are alike. For a short time after fledging, juvenile Spotted Flycatchers have pale spots on their wings and it's these that give the bird their name.

Listen out for short, shrill squeaks, particularly a *tek, tek* call. The most easily discernible is their alarm call, a sharp *eez-tek* or *eez-tek-tek*.

WHEN AND WHERE TO SEE

In parks, woodlands and large gardens from May to August.

Monthly musings

Gentlemen first

When it comes to birds returning to their breeding grounds, it is usually the males that are first back on the scene. Male Cuckoos, for example, start to arrive from late April, with the females arriving a little later. The males' eponymous call is both to mark their territory and a signal to potential mates. One reason put forward as to why males return first is that this gives them time to secure a territory in readiness for the females' arrival. However, researchers now think it could be a way in which the males maximise their chances to find a mate. Simply put, if you're already at the bar, you've more chance of spotting a date when potential partners turn up. Wait until the bar is heaving, and you'll have more trouble finding that suitable someone. The first-arriving females are also more likely to secure a high-quality mate, compared to those that leave it later.

Swallow stones

If you're up with your magical tales, you may have heard of 'Swallow stones', found in a bird's stomach and imbued with fantastical powers. The irony of this particular fantasy is that Swallows, unlike many other birds, do not eat stones. However, eating stones is not as strange as it may sound. Birds don't have teeth to chew their food but instead grind their food in their gizzards. To help the process, many birds, including garden birds such as House Sparrows, swallow grit or sand and hold it in their gizzard. This helps them break down hard foods such as seeds or nuts. One bird well known for its taste for grit is the Bearded Tit. These are a small wetland species with striking features that are often hidden away in reedbeds. In wetland nature reserves, wardens will often put out grit for them, aiding the birds' digestion and giving birders a rare chance to see these beauties.

← Male Cuckoo. ↑ Male Bearded Tit.

Top two must-dos

1 Watch an Osprey fish and feed its chicks

Ospreys are fish-eating birds of prey that migrate 5,000km from West Africa to breed in the UK. They are long-lived birds that return to the same nest site and will hook up with the same partner year after year, as long as they can ward off challenges from unmated birds. One pair, known as EJ and Odin, raised 17 chicks over seven years, nesting at RSPB Loch Garten in the Cairngorms in Scotland.

Back from the brink

Back in the 1800s Ospreys, like many birds of prey, were heavily persecuted, and the birds became extinct in the UK by the 1900s. However, in the 1950s a pair of Ospreys began breeding at Loch Garten and, determined to keep the birds safe, the RSPB immediately set up a 24-hour protection watch. Although the first breeding attempt failed, the Ospreys returned in subsequent years, going on to breed successfully for many years.

The Osprey's recovery as a breeding bird has been slow. Egg-collecting remained a constant threat and the use of DDT as a pesticide also negatively impacted on Ospreys. DDT in the food chain causes thinner eggshells in birds of prey, which dramatically reduces the birds' ability to breed successfully. Gradually, however, numbers increased, and more than 60 pairs were nesting in Scotland by the 1990s.

In 1996, Osprey expert Roy Dennis worked with the Leicestershire & Rutland Wildlife Trust, as well as Anglian Water, on a reintroduction project at Rutland Water in the Midlands. The first English-born Osprey chick fledged there in 2001, and Ospreys have been breeding at the site successfully ever since. Thanks to these and other pioneering efforts, Ospreys have spread to other areas and there are now hundreds of breeding pairs across the UK.

Ospreys' habit of building huge nests that are typically in isolated trees, plus their readiness to use human-built nesting platforms, also help make them easy to see, and a number of Wildlife Trusts and other organisations host pop-up information sites when the Ospreys are in town. Ospreys can be seen from March to September, but if you visit in June, you'll likely catch them feeding their chicks.

2 Hear a Cuckoo

Cuckoos could win the award for the most known about but least seen bird. Their familiar call was once a marker for spring, eulogised in poems and songs. Their outrageous nesting habits are well-known too, with 'cuckoo' a long-standing slang for crazy, and a cuckoo-in-the-nest being an imposter. But these birds are far less common now, being another summer migrant on the Red List of Birds of Conservation Concern. Yet to hear a Cuckoo is a thrill and it's well worth a trip to seek one out.

HOW TO SEE AN OSPREY

Going to see an Osprey nest has got to be one of the easier birdwatching activities out there, as often whole visitor centres have sprung up around them.

RSPB Loch Garten, Cairngorms
Visit the UK's oldest Osprey nesting site, with a state-of-the-art visitor centre and live cameras.

The Dyfi Osprey Project, Powys
Observe the Ospreys in season from the 360 observatory tower, less than 200m from the nest.

Rutland Water Join an Osprey cruise, giving you lakeside seats to see the birds fishing.

Watch at home Several well-known nests, including Loch Garten and Dyfi, are now watched over by live-streaming webcams, so you can see all the action online.

↖ An Osprey fishing is a thrilling sight.

→ Listen out for Cuckoos in spring.

Where and when

Cuckoos arrive in April, and right from the get-go the males sing their familiar two-note *cuck-oo*, announcing their location to incoming females. It is a clear two-note call, not to be confused with the three-note *coo-coo-coo* of a Collared Dove.

Cuckoos are widespread, found in woodlands, grasslands, wetlands and heathland – wherever their preferred host species nest. However, their numbers are much diminished, making hearing one more of a rarity than a certainty.

Identifying a Cuckoo

If you're lucky enough to hear one, the next challenge is to see it, as they tend to be shy and are often hidden away. A little larger than a Blackbird, the male has a grey head and wings, with grey stripes on its white belly. Females are similar but with a brownish tinge on the breast (a few are entirely red-brown). Juveniles are darker grey-brown and barred all over. When perched they often hold their tail half-raised and let their wings hang down.

A Cuckoo's stripy belly can cause it to be confused with a Sparrowhawk. In fact, back in the days when migration was a complete mystery, people believed that Cuckoos turned into Sparrowhawks for winter. Interestingly, Cuckoos are sometimes mobbed by small birds because of their raptor-like appearance. The thinking is that this enables a female Cuckoo to locate a target nest, as the birds leave their nests to give chase.

Fleeting visitors

Cuckoos spend most of their year in central Africa. Their time in the UK is short; they are here to breed and get out. The adults arrive from April and leave in June or July, with the newly fledged youngsters leaving soon afterwards.

It is the Cuckoo's notorious approach to parenting that enables the adults to leave so soon. Female Cuckoos steal into the nests of unsuspecting songbirds, laying a solitary egg for the much smaller host to rear. Females can lay up to 25 eggs in a season, most commonly in the nests of Dunnocks, Reed Warblers and Meadow Pipits. The Cuckoo chick will then evict any other chicks or eggs, ensuring that it has the undivided attention of the host birds.

← A Reed Warbler adult feeds a Cuckoo chick.

How to help

Give a little mud

House Martins and Swallows are known for their mud-cup nests. House Martins build closed-cup nests, building up to a beam or roof, leaving only a small hole from which they can fly in and out. Swallows, meanwhile, choose an open-cup design. Both use a mixture of mud and saliva, mixed with grasses or other plants to strengthen the structure. These mud cups are an ingenious design, but can prove tricky to build in prolonged spells of hot weather, because without water, mud can be very hard to come by.

There's a simple solution and that's to put out a bit of mud to give the birds a helping hand. Just leave out a dish of soil and water, and they'll make use of your muddy puddle should they need to. Alternatively you could create a patch of damp mud at the edge of a pond or garden border. The birds will be able to use the mud to help them build their nests, as well as to repair existing ones.

Peak building time is May. Both Swallows and House Martins will use old nests, repairing them as need be. Building a new nest can be laborious, taking one to two weeks and with the birds making more than 1,000 trips to collect material. It is also possible to buy or make artificial Swallow and House Martin nests and these can be a great way to help our summer visitors.

↑ House Martins make mud-cup nests.

Enjoy International Dawn Chorus Day

International Dawn Chorus Day takes place on the first Sunday in May and is a celebration of this seasonal wonder of nature when birds sing at their brightest and best. There's much research to suggest that birdsong can improve our feelings of wellbeing, so why not join the worldwide festivities?

International Dawn Chorus Day is timed for when birdsong is at its peak in the northern hemisphere, having built up during spring as resident birds are joined by summer migrants, all singing to attract a mate and mark their territories. The event is simply a call to get up early, head outside and revel in nature's symphony.

To hear the full dawn chorus you need to be up before the sun, as the singing peaks around half an hour before to half an hour after sunrise, which in May is around 5am. But whether you head outside, just throw open your window or decide to listen out for birds a little later in the day, it's a great way to mark the shifting seasons and connect with the natural world around you.

Both the RSPB and the Wildlife Trusts hold events, with experts on hand to tell you about the birds you can hear. Mindful of our human habits, you can also expect some 'not quite dawn chorus' events scheduled later during the day, with some even laying on brunch.

Taking time out to listen to birdsong might not seem like a way to help birds, but the evidence suggests that the more we engage with nature, the more likely we are to take action to protect it.

↓ Wrens are powerful singers.

Myth of the month

Beware a bird in the home!

Superstitions abound when it comes to birds and homes, with many cultures suggesting that a bird entering the house is a warning of bad luck or even death. Black birds, especially Carrion Crows or Ravens, in the house were considered a particularly bleak omen. An old Yorkshire belief warns that a Swallow entering your home down a chimney presages death, while in Norfolk a gathering of Swallows on your roof means death, as the birds would carry your spirit away when they fly back to Africa.

One explanation is that in times gone by, people placed great faith in a perceived natural order of things. Should this be broken, such as when a creature from the animal world enters the human one, something was badly wrong. The Norfolk belief adds a fanciful twist that plays into the birds' natural migration.

Birds nesting on our homes is thankfully a different matter and it is considered very lucky indeed if Swallows or House Martins choose to nest on your home. Conversely, if you harm either of these birds then bad luck is sure to befall you. This would seem to be a piece of folklore worth clinging on to. Both these birds are joyful summer visitors, captivating and entertaining in turn. When much of nature is sadly in decline, surely to be helping these birds by sharing a part of your home for their nests is a cause for celebration and delight.

↓ A Carrion Crow forages for food.

JUNE | 6

Floofed-up fledglings

This month we see the fruits of spring's hard labour. Young birds spill out everywhere, and suddenly the wider countryside turns into a nursery, with its family spats, misadventures and dramas.

IN JUNE, the largely private efforts of birds in and around their nests suddenly yield public results. All secrets are out, and chicks and fledglings make their way unsteadily into the world, prepared or otherwise. Young Starlings, looking oddly plain and brown in their juvenile plumage, beg loudly on lawns; the 'frothy' calls of young tits ring out from the verdant treetops; broods of ducklings or goslings hold up traffic. June's time of plenty is writ large.

For the birdwatcher, this can be hugely entertaining. There is something relatable about adult tits delivering food to their demanding, ungrateful offspring, who jostle each other and gobble down their food with juvenile uncouthness. Equally, who cannot root for the just-hatched chick taking uncertain steps, or for the little floating bumblebee that is a Coot chick? June is an involving month, which pulls at the heartstrings.

It is also a confusing month. All of a sudden, there are birds in front of you that don't fit their depictions in bird books and apps. Juveniles can be very difficult to identify. Youthful Robins lack their signature orange and look spotty; young Blackbirds are also speckled. Goldfinch juveniles lack any colour on their faces and wear a lost look. As for gulls, they become hotbeds of identification terror, as if they weren't already.

For birds, the tasks and responsibilities are onerous. Once away from the nest, their offspring are vulnerable in new ways, as they go wandering for the first time in a world full of predators and other hazards. And they may be just as hungry,

→ Juvenile Robins don't look like Robins without the orange breast.

← An Arctic Tern chick begs for food. June is a demanding month for parents.

but more mobile, so it is harder work to make sure everyone is fed.

However, to go further in appreciating the burdens of parenting on birds, we need to recognise a major distinction between two ways of bringing young up. These stem from how much growing up a bird does while inside the egg.

Helpless nestlings

The most familiar birds, including sparrows, tits, pigeons, crows and finches, tend to incubate eggs in the nest for about two weeks. When the young hatch from the egg they are blind, naked or with just a few clumps of down, unable to regulate their own body temperature and completely unable to find any food of their own. Termed altricial, such youngsters undergo their primary development in the nest, as 'nestlings', and only leave the nest when they are fully feathered, as 'fledglings'.

These birds are often mass-produced, with tits, for example, laying as many as 12 eggs, and many other common birds, such as Robins, producing six at a time. All those food packages delivered, all those caterpillars and flies, result in the sort of rapid growth spurt that we can barely imagine – within a fortnight a formerly helpless nestling is the same size as its parents, and able to fly. The plumage has the look of being rushed out, and so it is, offering cut price protection, and it won't be long until the youngsters moult again. As mentioned, most altricial youngsters have plumage that is distinctly different from an adult's, and usually more camouflaged; this might afford it some leeway from being attacked, but,

↑ These juvenile Kittiwakes are almost ready to leave the cliff-ledge nest.

→ An adult Coot feeds its young. Coot parents often 'split' the brood between them, each with assigned chicks.

equally as likely, its clothing identifies it as bottom in the birds' hierarchy, or pecking order.

Brood division

As a rule, both sexes feed fledglings, but the amount they do varies. In many cases, such as Blackbirds, the male takes the lion's share of early broods because the female is already committed to the next breeding attempt and may already be on eggs. In tits, the share is similar. In Chaffinch society, females do almost all the feeding.

The time spent feeding young out of the nest also varies. Young tits are independent within days. Crows may feed young for a month. Tawny Owls are also altricial, and they can still be working hard three months after the owlets have hatched.

After the period of feeding, the young quickly gain independence and will be ushered away from the parents' territory, especially if a new brood is in production. The youngsters disperse, often quite locally, although others, including Barn Owls, Little Egrets and Kingfishers, may move a lot further, travelling to different parts of the country or even abroad. Long-tailed Tits are an extremely rare exception. They remain with their families throughout the autumn and early winter.

Precocial young

The other type of youngster is known as 'precocial'. The eggs are incubated on the nest as usual, but they take longer to hatch out. Once they do, the chicks are in a fully functional state, able to run about and to take evasive action from danger, covered in warm down and often able to feed themselves. Such babies include ducklings and goslings, gamebirds such as Pheasants, and also waders and

Moorhens and Coots. As a general rule, they tend to be called chicks. These youngsters sometimes leave the nest within hours of hatching, never to return. That's why ducks and geese, for example, often wander around crossing roads and paths, a brood of babies trotting along behind, on their way to the nearest water.

The parental care of precocial chicks also varies. Some are fed by the adults for a long time. These include Moorhens and Coots but also grebes and several species of waders, including Oystercatchers and Snipe, who are given food that is hard for the chicks to find themselves. Others, however, can feed themselves almost from the start. Tiny, fluffy Avocet chicks look hilarious as they sweep their short but still uptilted bills from side to side.

↑ Avocet chicks are precocial, meaning that they hatch in an advanced state and can almost immediately feed themselves.

Other adaptations

Needless to say, nothing is simple in bird biology and there are variations on the above theme. Gull and tern chicks, for example, can run about like precocial chicks but don't leave the nest, as if they were altricial. They can freeze and hide, but trespassing next door could be fatal. They are termed semi-precocial.

Some nestlings, such as larks and buntings, are altricial but leave the nest before they are fully feathered or able to fly. These species nest on the ground, which is a perilous place.

To be honest, almost everywhere is a precarious place. At this time of year, Sparrowhawk chicks will also begin to hatch out. Their hatching lags behind many smaller birds, so that the predators can take advantage of the mass production of poorly plumaged, inexperienced, easily captured food.

Birds of the month

Grey Heron

THERE AREN'T TOO many birds that seem to spend much of their time standing still, but the Grey Heron does. The most typical sighting is of a bird with its feet in shallow water, looking down, waiting. Another familiar scenario is to see one just standing in a field or riverbank, hunched at the shoulders, giving off an undeniably grumpy air. Theirs is a slow life.

This familiar long-legged bird doesn't need to rush because it hunts for significant prizes. Its patient waiting by the waterside may well be rewarded when a fish swims idly towards its feet. The heron then lunges with cobra-like speed. A quick grab of a large fish can satisfy its needs for the rest of the day. The fish may wriggle, and an eel may wrap itself around the heron's head, but usually the struggle is in vain, and the prey slowly but surely slips down the hunter's long gullet.

At this time of the year, you see a lot of Grey Herons, because the young have just left the nest and are learning the fishing trade.

TOP ID TIPS

There is no other common tall, long-legged grey bird in most of the UK, so it's usually unmistakable. When it flies, the Grey Heron shows very obvious rounded wings, which form an arc, and it makes heavy flaps to make apparently ponderous progress. The neck is retracted and the long legs trail very obviously behind. The call is a loud, irritable *frank!*

WHEN AND WHERE TO SEE

Common all over the country throughout the year. It is found by ponds, lakes, slow rivers, canals, marshes and sheltered coasts, breeding in woodland.

Kingfisher

BRINGING A DASH of sparkle and glamour to the waterside, the Kingfisher is a great favourite of almost everybody and a sighting is guaranteed to make your day. Few birds in the UK are more dazzlingly colourful. The breast and belly are intensely orange, while the head, back, wings and tail shine metallic blue and turquoise. In the sun, the electric-blue back, in particular, seems to shimmer.

The Kingfisher's lifestyle is unusual, too. It lives up to its name, with great panache – who would have thought catching fish could be so exciting and wholehearted? It waits patiently on a perch just above sheltered water and, when it spots prey, it plunges in, head-first, in an attempt to grab the fish meal with its open bill. Sometimes it hovers over the water before making a plunge.

This isn't an easy way to live. The Kingfisher must first find a spot where it can see fish below it, in an undisturbed place – Kingfishers don't tolerate competition from rivals and are fiercely territorial. A hunting bird must compensate for refraction and must be accurate with its dive, otherwise the effort is wasted. If it isn't successful, it has no back-up plan – these small birds have no other way to feed.

To add to its hard work, the Kingfisher also digs a burrow in a bank for its nest, up to 1m long, which contains a nest chamber in which it may lay up to seven eggs.

TOP ID TIPS

The large head and bill and very short tail make an obvious profile, as it sits patiently on a spot overlooking water, be that an overhanging branch, a rock or even a bank. It often appears just as a blue flash over the water. It tends to fly low over the water, with rapid wingbeats. The call is a shrill whistle.

WHEN AND WHERE TO SEE

Quite common but it avoids uplands and doesn't like fast-flowing rivers. It can be seen in summer on slow rivers and lakes with suitable banks for nesting. It is much more widespread in winter, on all kinds of watercourses with still or slow water, and sometimes even on sheltered coasts, where it looks out of place by rockpools and mudflats.

Coot

IN SPRING our ponds and lakes can resemble battlegrounds, and this is in large part caused by the antics of that tetchiest of waterbirds, the Coot. A Coot in spring is cross with everyone and everything; individuals often charge over the water towards rivals – 'splattering attacks' – and approach others with an aggressive head-down posture, showing off the white frontal shield, which is larger in some birds than others and bigger in males than females. Coots also constantly call, uttering the curt *cut* which gives them their name, as well as various clucks and trumpets. They build a cup nest out of reeds and other vegetation, which is easy to see. Later in the summer, the chicks look like black bumblebees with legs, and have yellow tufts around their bald pink foreheads. Adults float on the water like ducks and usually dive for their food. They swim well and, out of water, you might notice their unusual feet, which are long, blue and with toes individually lobed, quite unlike the webs of a duck.

TOP ID TIPS

Once you realise it's not a duck, it's really easy to identify. The white frontal shield on the bill, plus the coal-black plumage, is unmistakable. Beware adolescent Coots, which have white fronts, are brownish and don't have the frontal shield.

WHEN AND WHERE TO SEE

Very common on larger bodies of freshwater, including ponds, lakes, broad rivers and marshes. Resident, but more numerous in winter.

↓ Coots fighting over territory.   ↑ 'Adolescent' Coots look very different to adults.

Moorhen

THE MOORHEN is a relative of the Coot (in the rail and crake family), with similar dark plumage. However, it doesn't form flocks on the water, preferring to walk around the edges of watercourses, often among reeds. When swimming it has a laboured action, head bobbing, like a cyclist struggling uphill. That's because it has very little webbing on its very long toes, which are mainly used for pattering over mud and clinging on to vegetation. It is just as hot-tempered in spring and summer as the Coot, and will also fight rivals with its feet, face to face on the water, kick-boxing with much splashing. It has a loud *kurruk* call, apparently surprised, as if somebody had just tickled it. Moorhens build very obvious nests, similar to those of Coots, comprising cups made from vegetation. One bird, usually the male, often brings material in, a delivery which is then placed on the structure by the female.

TOP ID TIPS

Easily told from the Coot by its bill colour, which is not white, but a waxy red-orange with a yellow tip. There is also an uneven white line along the middle of the body, and a pure white flash under the constantly jerking, pointed tail (either side of a black central stripe). The legs are green. Unlike Coots, Moorhens don't dive. Beware juveniles, which are grey-brown and lack the bright bill colour.

WHEN AND WHERE TO SEE

Ubiquitous all year round on almost any waterlogged habitat, from dense marshes to the edges of ponds and ditches.

↓ An adult Moorhen feeds its chick. The young look very different to the adults.

Great Crested Grebe

THIS IS A POPULAR and elegant aquatic bird of lakes and large rivers, easy to see and appreciate (see also page 35). It dives under the surface and chases fish, its principal food, back-propelled by its lobed feet and making use of its dagger-like bill, which is streamlined and ideal for grabbing slippery prey. It almost never comes to land, even building a nest of waterweed right on the edge of the water – the structure often floats. At this time of year there could well be a pair of Great Crested Grebes on your local lake, looking after their youngsters. Young grebes are stripy, as if wearing pyjamas, and they are very persistent in begging food. For the first 3-4 weeks of their life, they often rest on a parent's back.

↑ Grebes often carry their chicks on their backs, both on the nest and when swimming.

TOP ID TIPS

Easy to recognise, with its long neck and sharp bill. In the breeding season, this is a very handsome bird, with a 'double' wispy crest and fan-like cheek frills, which are tinged with chestnut. It looks very whitish in winter, with a black crown contrasting with white cheeks, and a white face and neck. It makes a distinctive braying call, which is often quite loud.

WHEN AND WHERE TO SEE

Mainly found on lakes and large rivers, so long as they have plenty of fringing vegetation. Prefers lowlands. In winter, this 'freshwater' bird often surprises people by inhabiting sheltered coasts, bobbing about on the sea, sometimes in big flocks.

June's challenge bird

Little Egret

THERE ARE STILL many birdwatchers in the UK who remember the Little Egret as a very rare bird; people would often travel miles to see one. It bred in Britain for the first time in 1996 on Brownsea Island, in Dorset. Now it is a fixture on the scene, especially in the south, with a population of about 1,000 pairs, and at least 11,500 wintering individuals. Its relatives the Cattle Egret and Great White Egret look set to follow it as conquerors of our country.

Nobody can say for sure why the Little Egret has exploded in numbers, but the best guess is that climate change, especially towards milder winters, has helped. Young egrets have a habit of dispersing significant distances (even across the Atlantic, for example), leading to large arrivals in late summer. If these survive the following winter, some are likely to stay to breed.

Little Egrets are colonial, and almost invariably join pre-existing colonies of Grey Herons in trees as add-ons. Their young make a hilarious gargling noise. The adults tend to feed alone, often along sheltered shores. Besides waiting patiently for a fish to appear at their feet, sometimes gently quivering one foot to attract fishy interest, Little Egrets are often seen dashing after prey, keeping their balance with spread wings.

TOP ID TIPS

Easy to identify as a pure white mini-heron, with a long and sharp black bill. The similar but rarer Cattle Egret has a shorter, thicker yellow bill and a thicker neck. Great White Egret is the size of a heron. The yellow feet of the Little Egret contrast with black legs; the colour is thought to help startle fish.

WHEN AND WHERE TO SEE

Common in the southern half of Britain and Ireland all year round. Easiest to see in coastal lagoons, but also occurs in freshwater lakes, marshes and meadows.

Monthly musings

Mallards and midges

Mallards are among the precocial group of birds (see page 97), in which the young hatch out in a well-developed state and can quickly run about and feed themselves. But what do you think ducklings eat in the first few days of life? The answer, surprisingly, is midges! In particular they feast on non-biting midges in the family Chironomidae, which are the ones that often settle on you when you go near rivers and ponds in summer. Ducklings often leap up towards midges in the air or settled on vegetation. At least two-thirds of their early diet is insect food.

↑ A duckling leaps out of the water to catch a flying insect.

↓ Parent Blue Tits often break the jaws of the caterpillars they bring to the nest.

Blue Tit burdens

If you have a nest box in your garden or nearby open space, you might have noticed the endless visits parent Blue Tits are paying to their hungry nestlings at the moment. Their burden is extraordinary, and they use all the daylight hours to provide for them. During the long period between eggs hatching and young leaving, which is 18–21 days, parents at their peak may have to bring in 500 caterpillars a day each, 1,000 in all. That's because each chick needs about 100 caterpillars a day, and there are usually about 10 chicks! Even if they only flew 10m there and back to find a caterpillar, that's 10km of flying.

Top two must-dos

1 Take part in a Peregrine city watch

It's easy to get despondent about the state of nature in the UK today. But look carefully and there are some significant good-news stories to enjoy. And one of those is the recovery here of one of the world's most charismatic birds, the Peregrine Falcon. In the 1960s the population dropped to 385 pairs, the decline caused by long-term persecution and by organochlorine pesticides like DDT, which caused eggshell thinning in the birds (see also page 133). Ever since that low, and especially since the banning of these chemicals, Peregrines have been increasing their numbers and range, with almost 1,750 pairs in the UK today.

Perhaps what nobody expected is that Peregrines would thrive away from wild and remote places such as mountains and cliffs. But that is exactly what they have done. Surprisingly, they have found refuge in the hearts of many of our

larger towns and cities, where they have access to two critical resources. Firstly, the many tall buildings make safe and secure places to nest. Secondly, they have easy access to one of their very favourite foods, pigeons.

The result is that urban dwellers can come face to face with one of the most thrilling predators in the world. The Peregrine is the fastest-moving animal on the planet. It relies on speed to catch and disable its avian prey in mid-air, often striking it with its talons and producing an explosion of feathers. When diving down from a height, it can reach a barely believable 389kph in freefall. It easily breaks 80kph in level flight, and its

← Peregrines choose a high spot for their nests, ideal for spotting potential prey.

astonishing control and acceleration is quite something to witness. And you can – on a shopping trip!

In June, Peregrines are hard at work bringing in prey to their young. The nests are often on the ledges of large buildings, such as cathedrals or hospitals, well above the hubbub of busy streets. The birds are so busy that, in some places, the RSPB or Wildlife Trusts have set up Peregrine Watch telescopes to allow city people to enjoy a close-up view of a wild bird tending the nest. Better still, it will bring in the latest feathered offering to the hungry chicks. Many people have no idea that this enthralling drama is going on in the skies above them. Lots of visitors, once they know it is happening, tune into one of the many webcams that are trained on the eyries throughout the season.

There are now so many urban Peregrines (there are more than 20 pairs in London alone) that it would be impossible to list all of their locations here. And that is, in itself, evidence of a heartening recovery.

2 Listen to a Nightjar

Is there anything better than June evenings in the UK? Yes, there is – June evenings involving a Nightjar adventure. One of our most extraordinary birds, the nocturnal Nightjar plies the summer skies for insects such as moths and beetles, catching them by sight during the twilight and the night. It is the only British bird known to coordinate its breeding cycles with the phases of the moon. It prefers a full moon when it is feeding its newly hatched young, when the extra light helps it to catch prey. A strange, large-headed bird, it has hardly any bill at all, but its enormous, touch-sensitive gape

← A Nightjar alights on a favourite song post.

Somehow somebody managed to besmirch the Nightjar's reputation by suggesting that it had a thirst for sucking the teats of goats and calves and causing them harm. This idiotic notion is retained in the bird's scientific name *Caprimulgus* ('goat-sucker').

To hear a Nightjar, you must venture into its habitat at dusk (at this time of the year, of course, that can be close to 10pm). Over much of the country, that habitat is lowland heathland, although the bird also occurs on the edge of moors and in woodland clearings. Either way, there is nothing to

and fringes of mouth bristles enable it to snatch and hold what it intercepts in the evening air. By day, the Nightjar uses its A-grade camouflage to disappear completely against the ground litter, or against the branches of trees. It is all but impossible to find one at rest.

The Nightjar would barely register with birdwatchers if it wasn't for its quite extraordinary, other-worldly advertising call. It is completely different from every other bird vocalisation you might hear in the UK. Described as 'churring', it is a long drilling sound with two obvious 'gears', a sort of wooden trill which is entirely distinctive. The fact that it is heard at night adds to its allure and strangeness.

Our forebears couldn't cope with the eerie voice, nor its unseen source.

beat wild places in late evening. Your senses heighten and you become aware of the smell of heather and gorse, of the whisper of the slightest sounds, and of the feel of the first mosquito bite! The Nightjars don't sing until it is nearly dark, and the moment you hear them there is always a flutter of excitement, no matter how many times you have heard it before. If you are lucky, you might catch a glimpse of the bird's aerodynamic form in the darkness, but this is never guaranteed.

It is safer and more fun to go Nightjar-seeking with a group of like-minded people. There is something slightly crazy about going into the darkness to hear a bird, and it's a real adventure away from your comfort zone in the zenith of summer.

How to help

Record your sightings with Swift Mapper

How about this for a scrumptious way to help conservation? You choose a glorious warm June morning and go for a walk in your local area, perhaps to a few places you've never explored before. It could be a small village, or a part of town new to you, and it's fine for a coffee shop to be involved. Your mission is to watch out for Swifts, those airborne summer screamers, the quintessence of Britain's fairest season as they scour the sky for small insects wafting in the breeze. Honestly, can you think of a more pleasant way to pass the time?

And if you see Swifts, either in their distinctive 'screaming parties' or flying up towards potential nest sites, you can log your sightings on the Swift Mapper app with a few touches of the screen (don't log Swifts just flying around, because they range widely from breeding sites). That's it – you have done your bit to help this wonderful bird.

Up in the rooftops Swifts live a secretive life, above human reach and away from prying eyes. They are completely dependent on their insect diet, but also toy with the uncertainties. The female adapts the clutch size to the current food supply, laying one, two or three eggs. If, after laying, there isn't enough food, the adults don't risk their own health but stop feeding the chicks altogether. If a large weather system passes by during the breeding season, the adults sometimes evacuate right away, some travelling many hundreds of kilometres, to find better conditions. To cope with the adults' absence during these evacuations, young Swifts are strongly resistant to starvation. Nevertheless, however adept these birds might be at flying, their success is always balanced on a high wire.

↓ A screaming party of Swifts sweeps over rooftops.

Swifts, as mentioned on page 81, are in steep decline in Britain and Ireland – we've lost over half our breeding pairs since 1990. One of the reasons is that colonies are being inadvertently disturbed or destroyed, and the number of places where Swifts breed is fast diminishing. In order to alleviate this, conservationists need information.

Backstreet gangs

Swifts are not the sort of birds that can be looked after in nature reserves. They are extremely widespread and most of the colonies are on buildings (a few are on sea cliffs). The colonies are rarely large, and hundreds are tucked away in places where birdwatchers seldom go with binoculars, such as housing estates and town centres. Swifts are drawn especially to old buildings, and particularly taller ones that allow at least 5m of clear space to fly to and from their ledge-based nests.

In order to protect these precious populations of Swifts, we need to know where they are. That way, if a housing development or big renovation is planned for a town centre, local conservation organisations can advise contractors about ways to accommodate a bird that is legally protected. Whether large or small, protecting every colony is vital.

If Swifts are breeding somewhere, it is easier to expand a colony than to create a completely new one. Again, if we know where they are, conservationists can advise where to put special nest boxes called 'Swift Bricks' into new builds to tempt the birds to form a satellite colony.

So, download the Swift Mapper app and join your local Swift group, too.

↓ Artificial nest boxes are of immense help to Swifts.

Myth of the month

The mystery of the baby pigeon

It's become a modern piece of folklore, especially for those who live in cities throughout the world. Why do we never see baby pigeons? There are numerous websites and publications which have elevated this simple query to one of Life's Big Questions.

The answer is very simple and is just a matter of perception. We don't see baby pigeons because we don't recognise them for what they are.

As described on page 96, birds have different ways of parenting young, and pigeons are firmly on the altricial spectrum. In other words, when the eggs hatch, the young are not ready to leave the nest. In fact, it takes them a long time to leave – no less than 35 days. During this time they are nourished by the 'crop-milk' of the adults – a nourishing stomach secretion. By the time they fledge, they are so well developed that not only are they the same size as the adults, they hardly look any different, just a little duller and paler. That's why we don't easily recognise them.

To be fair, the young pigeons (squabs) actually in the nest are very difficult to see. For example, Feral Pigeons nest on cliffs, on ledges on precipitous buildings, and on bridges and roofs, often high above street level and both out of sight and out of mind of most inhabitants. Because we rarely see the squabs, adult pigeon seems to beget adult pigeon. It's an urban myth.

↓ Apart from looking ragged, juvenile pigeons, like this Wood Pigeon, are very similar to adults.

JULY

7

Summer by the sea

For some birds, the breeding season draws to a close this month, and they begin to take stock, recover their energies and moult their feathers. Many others, including seabirds, are still in the thick of reproduction.

JULY CAN APPEAR a quiet month for birds in town and country. The breeding frenzy is largely over, and exhausted parents are finally attending to a little self-care. Many, such as your beloved garden Robin, begin to moult their feathers and settle for a low profile. Replacing feathers is hard work, and many birds become much less obvious, especially in the garden, where householders sometimes get worried about them.

Some birds are still as evident as ever. If you care to look, pigeons are still displaying vigorously, their crooning

songs and rooftop song-flight very much a part of the suburban summer, not that they get much of an appreciative audience. Yellowhammers and several finches, especially Goldfinches, are still breeding, the former because it is peak season for their second-brood nestlings' summer food, grasshoppers, and the latter because of the late summer seed crop, especially of thistles. Some Blackbirds and Robins are also still tending a third brood.

It might be summer, but many birds have already begun their travels. Adult Cuckoos have no reason to remain and leave quietly; some Swifts have gone by the month's end. Furthermore, some

← Puffins are unmistakable.

↓ A Yellowhammer may sing 7,000 times a day.

failed breeders from further north, such as Green Sandpipers and Whimbrels, have already turned up on muddy corners and coasts on their journey south.

But it's still hard not to get the impression of quietness, of a party largely over, the dancefloor messy but largely deserted. However, there is one place where this is not the case at all – by the sea, where all the hustle, bustle, intrigue and din of reproduction is still raging.

Special seabirds

In a world context, Britain's seabirds are the most significant part of our avifauna, with waders not far behind. There are several species for which we hold large parts of the world population. For example, we have 80 per cent of the world population of Manx Shearwaters, 58 per cent of the world's Gannets and 60 per cent of the world's Great Skuas. We host vast colonies of Puffins, Razorbills, European and Leach's Storm-petrels, Kittiwakes and various terns. Much as we love our Robins and Goldfinches, these are insignificant in a world context. But our seabirds truly rock.

The British Isles lie, of course, in the North Atlantic, a vast resource for seabirds. We have an exceptionally long coastline, many islands and islets and tall cliffs. Seabirds spend most of their lives at sea, but in the breeding season they have to find safe landfall, where they can lay eggs out of reach of as many land-based predators as possible. For many seabirds, that means precipitous cliffs and sea stacks, but it can also mean low-lying islands and dunes.

There are never enough of such places to go around, and this means that many seabirds are densely colonial, packing in as many breeding birds as possible. Seabird colonies also need access to the sea, which is where all their food comes from, including anything they are going to feed to their young. The wonder of crowded sea cliffs is a result of this dual need for safety and access.

Comings and goings

A seabird colony throbs with noise, movement and energy, and can be a confusing sight. But much can be explained by the fact that all the birds present need to commute to and from the sea to get food for themselves and/or their chicks, and those that aren't commuting are either looking after the nest, eggs or young; are unable or not ready to breed or have lost their nests (hence are lounging around); or are bickering with each other.

For some bird colonies, especially those of terns, for instance, foraging tends to be close at hand in the shallow waters nearby. For most seabirds, though, travelling for provisions requires a substantial journey. It is fair to assume that your average Razorbill or Guillemot, for instance, commutes at least 25km out to sea every time, making a 50km round trip. Many seabirds travel a great deal further than that: the Gannet is one example of a bird that will go at least 400km to obtain food, and it has been recently confirmed that Leach's Storm-petrels go foraging as far as the Mid-Atlantic Ridge. Most seabirds only make a handful of visits to their young each day – six in the case of the Puffin – and many young seabirds have a degree of built-in resistance to hunger. Nevertheless, not everybody travels at the same time, so any seabird colony is a hub of constant comings and goings.

At some seabird colonies, a very different timetable is followed. Manx Shearwaters and both European and Leach's Storm-petrels only visit their nest holes by night. This applies both to pre-breeding, when the noise emanating from birds howling or caterwauling underground in order to attract a mate must be heard to be believed, and to the later part of the season when they are provisioning their young. These species all nest in crevices and burrows and pay nocturnal visits to avoid predation.

↖ Guillemots often have to make a long journey out to sea to find food for their young.

→ Manx Shearwaters only visit their breeding colonies at night, to avoid predation.

Running the gauntlet

In their lives, birds never stray from the shadow of predation, and the seabird colonies, densely packed with vulnerable eggs and chicks and their busy and often distracted parents, overflow with opportunity for predators and thieves. The main dangers are from gulls and skuas, which patrol colonies on the lookout and are rarely disappointed. In some areas, Peregrines and White-tailed Eagles take their toll, and the adult sea birds are not necessarily safe either. Many individuals specialise on specific areas of cliff, or even on certain species. Some Great Skuas in the Western Isles have taken to feeding at night on storm-petrels. Thievery is a constant threat, too. A 50km round trip can end with everything being stolen by a gull or, especially, a skua. Some Arctic Skuas depend on stealing food from other seabirds to ensure their breeding success.

Despite the dangers, many seabirds are much longer-lived than land birds. A Puffin lives an average of at least 20 years, while a comparably sized pigeon only manages three. Year after year, many seabirds breed with the same mate, at the same site in the same colony. Their long lives mean

that their reproduction can be very slow. Gannets, Guillemots, Puffins and Manx Shearwaters, for example, lay only one egg a year, and may not breed every year – each breeding effort entails a heavy investment in that one chick, and pairs only attempt to breed when they are both in peak condition. If disaster strikes, such as with the recent (2020s) outbreak of bird flu, this can make their populations highly vulnerable, as they cannot quickly 'bounce back' from heavy losses.

→ Great Skuas often harass Gannets when the latter are returning to the colonies with food.

Birds of the month

Black-headed Gull

IT NEVER HAS a black head! However, you can't miss the smart brown hood in the breeding season, which diminishes to a dot behind the eye in the winter. This diminutive, elegant gull is the commonest and most widespread gull in Britain in the winter, when it is everywhere. In July, most birds are still at their breeding colonies, which are on coasts, on islands in lakes and marshes, and on moorland. The colonies consist of fairly large platform nests made of vegetation, into which three eggs are laid. The birds never stop making their loud, rolling calls, which always seem to be unnecessarily excitable. It is highly adaptable, and is often seen feeding behind a plough, fighting with ducks for grain and, in late summer, consuming flying ants and other airborne invertebrates.

TOP ID TIPS

A small gull with slender wings which have a sharp tip. The chocolate-brown hood, together with the white eye-ring, make it unmistakable in breeding plumage (but beware the rare but increasing Mediterranean Gull, which does have a black head). In contrast to our other numerous gulls, the bill is either red or orange, and brightest in winter. Young birds have some brown mottling on the wings. In flight, the front edge of the wing shows a long white wedge. It has many calls, such as *kek* and a rolling *kree-aah!*

WHEN AND WHERE TO SEE

An abundant gull, especially in winter when it is numerous inland. Found throughout the country.

↙ In winter, the Black-headed Gull has a smudge behind the eye. In summer, it has a smart chocolate-brown hood (inset).

Herring Gull

THE CRY OF GULLS is the quintessential sound of the seaside, and throughout most of the UK, the 'seagull' is the Herring Gull. This is the one that yells from the rooftops of coastal towns, that gathers in flocks on the beach and often just seems to hang around. It is one of life's opportunists, perfectly content to rummage around bins and steal chips from tourists. At the same time, it is a seabird, able to catch fish from the surface, ride big waves and sail through storms with its wondrous, effortless flight. It is an inveterate stealer of food, both from its own kind and from other birds. It is also often seen on the coast dropping seashells from a height in an attempt to break them open, showing the intelligence that underpins its adaptability.

TOP ID TIPS

If you see a big gull with a white head and silvery-grey back and wings, plus white-spotted black wingtips, it is almost certainly an adult Herring Gull. This species also has pink legs, plus a yellow bill with an orange mark. The adults have yellow eyes and a distinctly angry expression. All gull flocks tend to contain a lot of young birds with their mottled brown plumage.

WHEN AND WHERE TO SEE

It's abundant all around our coasts, occurring everywhere from harbours to big sea cliffs. It breeds in some inland big cities on rooftops. In winter it can be found inland on lakes and reservoirs and by rubbish tips.

↓ An adult Herring Gull (left) feeds its brown-speckled offspring.

Kittiwake

THIS IS A GULL, but not as we know it. It is an ocean-going species that keeps itself to itself, and that means usually well away from people. It primarily feeds on fish, which it snatches from the surface out at sea, often far from land. In the breeding season, however, it congregates on cliffs, often in huge colonies, mixing with Guillemots on the narrowest ledges. In contrast to these neighbours, which lay and incubate their solitary egg on the bare rock, the Kittiwake makes a nest out of mud and vegetation, often seaweed, lodged precariously. Pairs routinely steal nest material from their neighbours. This noisy bird forms the soprano section of the seabird cacophony, making a loud series of wails – *kitti-WAAAKE* – that are easily heard above the rumble of waves and the grunts of other seabirds.

TOP ID TIPS

It looks like a gull fresh from a makeover and a gym session, perfectly turned out with a gleaming white head and tail and smart grey wings, which have black tips as if dipped in ink. In winter they develop a dark head spot and a grey 'neck boa'. The dark eyes have a gentle expression and it has a yellow bill. The legs are noticeably short and black, unlike those of other gulls. Immatures are not mottled brown like most other young gulls, but are smartly patterned with black zig-zags across the wings.

WHEN AND WHERE TO SEE

It breeds all around Britain and Ireland where there are cliffs, but this makes it scarce in the south and east. It is abundant in northern Scotland in particular. After breeding it flies out to sea and may reach North America. Storms bring birds closer inshore in the autumn and winter.

Common Tern

TERNS LOOK RATHER like gulls, but are smaller and more delicate, with longer, forked tails, sharp bills and sharp wingtips. Unlike gulls, they have short legs and don't walk around much. Despite being seabirds, they don't swim (although they can). They are just as noisy as gulls – indeed, if you can believe it, even noisier – with sharper, less clanging calls. The Common Tern is our most widespread species and the only one that breeds inland away from northern Scotland, often on lakes, especially if those are provided with anchored, floating platforms called 'tern rafts'. In common with most terns, it hovers on the spot over water and dives down for fish, and sometimes insect larvae. Terns nest in colonies and often lose their cool, panicking in the presence of enemies visible and invisible.

The comings and goings are fun to watch, because birds can often be seen flying in with fish in their bills.

TOP ID TIPS

Quite a grey-looking tern, with red legs and a red bill with a black tip. As with all our breeding terns, it has a black cap. Among the various calls are an impatient-sounding *kek* and a drawn out *kree-arr*. Beware the very similar Arctic Tern (see page 123), which has an entirely red bill, shorter and blunter than the Common Tern's, with longer wings and tail.

WHEN AND WHERE TO SEE

A widespread summer visitor, from April to September, breeding on islands and beaches, especially with shingle. Mainly coastal, but also breeds inland.

Sandwich Tern

THE SANDWICH TERN is bigger than our other terns and its pure white and very pale grey plumage is distinctive at a distance. More powerful than its relatives, it is often seen flying higher and with laboured-looking wingbeats, working its way along above the sea, with its head distinctively pointing down, peering at the water surface for fish. When it dives down to catch one, it plunges fully into the water and makes an obvious splash. The Sandwich Tern breeds in large, noisy colonies on shingle and sandbanks and is a nightmare for conservationists because the colony often shifts site at the beginning of the season, for no obvious reason. Young terns gather into creches for safety, but still beg when their parents come in with food. Some remain dependent even at the start of their migration to West Africa.

TOP ID TIPS

Our biggest tern, about the size of a Black-headed Gull, but much slimmer, with a sharp bill and wingtips and a longer, forked tail. It is our only tern with black legs. The bill is black with a mustard-yellow tip (the mustard in the Sandwich, if you like). It has whiter plumage than other terns, a shorter tail than most, and, in summer, a shaggy crest. It has a very distinctive *kerr-ick* call.

WHEN AND WHERE TO SEE

A summer visitor to coasts between March and October; a few remain in the very far south for the winter. Widespread but has only a few regular breeding sites.

July's challenge bird

Chough

THE CHOUGH IS A CROW, but an outlier of a crow. It tends not to do those 'crowish' things we dislike, such as cawing loudly, eating cute vertebrates (not while we're looking, anyway) and loafing around rubbish bins. This is the refined member of the family, with its signature red downcurved bill restricting it to probing in soil, mainly for earthworms, ants, flies and other invertebrates. For now, we'll overlook its habit of probing into dung. The Chough is a rare bird, found in a few scattered, almost invariably coastal localities where it usually nests in sea caves, quarries or even mine shafts. One of this bird's interesting quirks is that breeding pairs are sometimes helped at the nest by progeny from a previous year.

The Chough has a delightful, carefree demeanour, often performing aerobatic manoeuvres in flight, typically with colleagues in the flock, and giving its friendly, cheerful call. It often closes its wings in flight and dives down. It's one of those birds that, once you've seen one, is quite easy to identify just by its character and demeanour.

TOP ID TIPS

Easy to identify if seen well, because the red bill and legs are unmistakable. Young birds have yellower bills. In flight it shows rounded wings with very obvious 'fingers' at the tips. The call is *cheeow*, like a higher-pitched version of the Jackdaw's *jack* note.

WHEN AND WHERE TO SEE

Resident year-round in a few choice localities, including Cornwall, parts of Wales, the Isle of Man, Islay and Northern Ireland. It is being reintroduced to south-east Kent and some other places.

Monthly musings

The Arctic Tern's migration

The Arctic Tern, a bird that breeds in the UK's coastal areas, especially in northern Scotland and Ireland, is famous for having the longest back and forth migratory journey of any bird in the world. It breeds in northern latitudes, including the Arctic, and spends the non-breeding season in the Antarctic, a minimum distance of 19,000km. However, it doesn't necessarily go directly. Tracked birds from Iceland may travel up to 70,000km a year and Dutch birds on average travel 48,700km. The longest recorded so far is 81,000km. These birds see more daylight in their lifetime than any other living organisms.

Anting

One of the strangest pieces of bird behaviour that you can ever see is the practice of anting. There are two types, observed in a minority of bird species. In passive anting, birds crouch by a nest of ants, spread their wings and tail and allow ants to crawl over them. In active anting, birds collect ants in their bills and smear them all over their plumage. Although unusual, anting presumably aids in plumage care. The ants may attack lice or other ectoparasites, or their formic acid, when applied to the feathers, may cause the lice to move and be more easily preened off. Birds sometimes 'ant' using millipedes instead, which also possess noxious chemicals.

↑ Arctic Terns have the longest migratory journey of any bird.

← A Jay allows ants to crawl over its plumage.

Top two must-dos

1 Visit a seabird city

To visit a big cliff-based seabird colony at the height of the breeding season makes for an enthralling experience. It hits many of the senses. The sheer overwhelming crowds of moving birds assail the vision, and the smell of seabird guano assaults the nose. And the noise, the noise! In a big colony, the Guillemots make an ululating trumpeting, the Razorbills make a rattle like a ship's timbers groaning, while Puffins sound like motorbike engines being tuned up. Gannets sound like they are roaring down a long cane, while Fulmars cackle like a coven of mischievous witches. Against it all, the Kittiwakes squeal as if belting out on a brass instrument. At night, storm-petrels and Manx Shearwaters sound like demented hordes of lost souls, and, as the listener, you might begin to doubt your own sanity.

There is order, though, because everybody has their place. The Guillemots are the base-jumpers flirting with death, choosing absurdly narrow ledges and nesting so close that they are touching – their colonies are the most densely packed of any bird in the world. Razorbills take the broader ledges, often with an overhang, or within a small crevice, and have room to move around. Puffins nest in rock cavities, completely hidden to the outside world; where there is clifftop soil, they will dig burrows. Kittiwakes mix with Guillemots on thin ledges,

→ A Fulmar rides the clifftop breezes with effortless ease.

↙ A crowded colony with Guillemots, Razorbills and Kittiwakes.

but they build a nest anywhere they can lodge mud and plant material. Fulmars and Herring Gulls take the penthouse suites – broader ledges, often near the clifftop. The biggest birds, such as Gannets and Cormorants, take the widest ledges; Shags often take to sea caves close to the water.

Every one of them comes and goes. The Razorbills and Guillemots take off from their ledges and fly low over the sea with quick wingbeats. By contrast, Fulmars and Herring Gulls wheel effortlessly along the clifftops, riding every gust of wind expertly, like albatross tribute acts. Gannets trundle off their ledges and use the gusts to get airborne, while Cormorants fly low to the waves and seemingly cannot wait to get into the water.

A seabird colony is a thousand soap operas in one, although the dramas you'll see are real. Watch carefully and you will notice neighbours bickering, pairs greeting, parents providing food and young being sheltered. You will also see a surprising amount of downtime, with many birds seemingly doing nothing at all. Equally, while you are there, a family could suffer predation – a breeding season ruined – or a chick could receive a much-needed meal. It's a cliché, but all life is here. It will stay with you.

2 Look for a bird sunning itself

It's July, so we are entitled to expect some fine weather. If that happens, people will be out in droves, stripping off and worshipping the sun. You might not realise it, but birds also sunbathe, and give every impression that they enjoy it as much as we do.

As with many subtle pieces of bird behaviour, sunning is best observed in the garden. The most likely participant is the Blackbird. Watch for a bird resting on the ground with all its feathers ruffled and its wings and tail spread. It will have its bill open, too, which helps it to lose heat; birds don't have sweat glands. A sunning bird can remain in position for minutes on end, looking almost stupefied. Pigeons often sunbathe while leaning slightly to

← A Grey Heron in its typical sunning posture.

↓ A Blackbird sunning itself.

one side, with one wing open and the other closed (they assume the same posture to bathe in a rain shower, too).

Being out in the open in such a posture makes a bird vulnerable, so we can be quite sure that there is a good reason for doing it. Scientists have long scratched their heads about what that might be, though. The most plausible is that it helps birds remove ectoparasites, such as lice and ticks. A bird's plumage may be plagued by these, and there is evidence that high infestations make a bird less attractive to a potential mate. Recent studies have shown that lice can be killed by even a short exposure to direct sunlight; those from a group of vultures succumbed after 10 minutes when the sun heated them to 60°C. The UK's milder temperatures could still do an effective job. In another experiment, North American Violet-green Swallows were sprayed with bird-safe pesticides and spent considerably less time sunning than the birds in the control group.

Even if the sun doesn't kill parasites directly, being exposed might force them to relocate, in which case they might come within reach for a preening bird. Many lice are very small, just 1mm or so long, and spotting them can be difficult. If they are moving, that might make things easier. To support this theory, it is known that birds almost invariably preen after they have been sunning.

There are other theories. Direct sunlight could kill harmful bacteria that compromise feathers. And sunlight can act on the keratin of feathers to stiffen them.

Whatever the reason, seeing birds sunbathe is a fascinating sight. And of all our birds that indulge, the Grey Heron is the most comical sight of all. It spreads its wings and droops them down, looking for all the world as though it is about to catch a large ball.

How to help

Fairs, festivals and fêtes

It is the fair season, and the season of fairs. Summer is the time when, all over the country, people organise myriad local fêtes where neighbours declare their colours and interests by helping out at stalls. It is also a time of festivals and fairs, including birdwatching's 'Glastonbury', the Global Birdfair (formerly British Birdwatching Fair), held in Rutland. All of these are wonderful opportunities to help nature, by doing anything from paying an entrance fee to attend an event, to raising money for charity, selling bird food, or just talking to people about your love of birds and wildlife.

If your local event doesn't have a stall related to wildlife, why not get involved and see if you can set up one yourself? The worst thing that can happen is that you eat a lot of cakes.

Volunteering

Nature always needs a helping hand, and if you would like to contribute to conservation beyond your garden or usual areas of influence, there are countless ways to contribute. It hardly needs to be said that volunteering for wildlife brings enormous benefits, because there is an increasing body of experience and

↓ RSPB staff and volunteers are often on hand to help at fairs and fêtes.

research that tells us that being outdoors is good for us in every way, both mentally and physically. You also benefit from being in the company of other people and out of your bubble. It could even help you in your career.

There are many conservation charities and all, invariably, need help. It's a cliché, but anything will do, from a day to a regular commitment. The RSPB offers internships or residential stays on its reserves, where volunteers can help with practical everyday tasks, from helping clear vegetation to manning the shop. Many charities organise conservation tasks. Organisations are particularly keen for volunteers to help them with member recruitment, one of the less 'sexy' but still vital tasks that need doing. The RSPB has a part of its website which seeks

to match willing helpers with the many tasks that require bodies. You may also find local projects that centre around improving habitats for wildlife on nature reserves and in public spaces, or other green initiatives such as litter-picking and beach-cleaning.

Another area where volunteers are always needed is in Citizen Science projects, of which the Big Garden Birdwatch (see page 19) is one. These are vital for monitoring wildlife, but rarely need a major time commitment, just reliability. There is a project for everyone, from counting butterflies to reporting sightings of Stag Beetles. If you have time to spare, you won't have the slightest trouble filling it.

↑ Volunteers ready for action to help wildlife.

Myth of the month

Ravens and the Tower

There is always a lot to worry about when it comes to the state of Britain in the twenty-first century. But we shouldn't be fundamentally concerned. Our fate rests not on the whims of kings or politicians, nor on the threat of foreign powers, but on the state of a small group of captive Ravens kept at the Tower of London. According to legend, if the Ravens are lost or they fly away, the Crown and the kingdom with it will fall.

Of course, it is difficult to prove or disprove this legend, without actually taking the Ravens away. And that would be too risky, because we don't actually want the worst to happen. It is better to make sure that the birds are well looked after, and there is, happily, a Yeoman Warder Ravenmaster there to do just that.

Nobody seems to be quite sure where the idea of keeping Ravens at the Tower comes from, although there are claims from the reign of Charles II (1660–1685), and he was just the sort of king to do something original like that. Equally, some say it dates from as recently as the Victorian era.

The folklore surrounding it could go much further back, to Celtic tradition. In Welsh mythology, Brân the Blessed was a king of Britain and his severed head is buried under the Tower of London so that his skull can watch over the kingdom. The name Brân means raven or crow, so it is possible that the tradition of keeping Ravens stems from this association.

→ A Raven perched by the Tower of London – all is well.

AUGUST

Hiding and gliding

Hot summer days and all the birds are up, up or away. Soaring high on the thermals are mighty birds of prey, but our parks and gardens can seem strangely bereft of birds. What's going on?

AUGUST CAN SEEM like a very quiet time for birds. The feeders barely attract a visitor, the early morning dawn chorus seems a distant memory and there just isn't that buzz of avian activity when you're out and about. It's all a bit strange, and among the most common calls to the RSPB's Wildlife Enquiries Team at this time of year are those from worried bird-lovers, concerned that all their garden birds have mysteriously disappeared.

All change

Thankfully there's a very good reason. Our common garden birds haven't so much disappeared as gone on their annual retreat. A bird's feathers, much like our favourite outfits, eventually wear out over time. Birds deal with this by moulting to replace these frayed and faded feathers with new ones. But as the old feathers fall out and the new ones grow, a bird is vulnerable. Birds need their full and proper plumage both to fly and to regulate their temperature,

helping them keep warm when cold and cool down when hot. But during moult, their ability to fly and regulate their temperature are both affected by the feather loss. The safest way to manage the moult is to hide away. It's a far cry from the bold business of territorial singing and the frantic work of raising chicks, and this makes our parks and gardens feel all the quieter. This year's chicks will also be experiencing a moult, as they swap their soft juvenile feathers for more robust adult or subadult plumage.

Many birds don't moult into the rich, showy plumage sported in the spring breeding season, but become slightly more muted versions of themselves. We'd still recognise them as a Chaffinch or a House

← Look out for Ospreys over water in August.

↑ During the moult, birds, like this Robin, can look scruffy!

131

Sparrow, for example, but it's not until next spring when the breeding season starts again that we can expect the birds to look their smartest, as the new feathers have dull-coloured fringes that need to wear away to reveal more of the brightly coloured parts. It's a bit like keeping your best outfit for a special occasion.

Take my breath away

One group of birds that are often visible at this time of year are the birds of prey. As the summer sun beats down, it's the ideal time to look up for raptors. Many birds of prey, such as Buzzards and Red Kites, take advantage of thermals, warmer currents of rising air. By catching the rising air with their wings, they can soar, minimising the need to flap their wings and saving their energy.

Characterised by their fearsome talons and flesh-ripping hooked bills, raptors are supreme hunters, some of them occupying the very top of the food chain, while others mainly scavenge for food. They have sharp eyesight to pinpoint prey from a distance, and are masters of flight. We have 15 breeding birds of prey in the UK, from large and powerful eagles and buzzards to the smaller fast-flying falcons. We also have five species of owls, but the term 'raptors' is usually reserved for daytime flyers.

Many of our 15 raptors are resident in the UK all year round and these include Red Kites, Buzzards, Peregrines and Kestrels. Others such as Ospreys and Hobbies migrate to the UK in spring and can still be seen in August. Another summer migrant is the Marsh Harrier. This is a large raptor that is found in wetlands, often seen hunting over reedbeds, with the majority found in the east and south-east of England. More recently, however, more and more Marsh Harriers have been overwintering in the UK, rather than returning to Africa.

↙ Carrion Crows regularly mob raptors such as Buzzards.

↑ Marsh Harrier hunting over a reedbed.

The fall and rise of raptors

Birds of prey have an uneasy history. In the nineteenth century, the rise of game shooting, together with a fashion for egg-collecting, saw numbers decline, and by the end of the First World War, five of our 15 native birds of prey had been driven to extinction in the UK. The advent of pesticides in the 1950s and 1960s also hit raptors hard. A group of poisons known as organochlorines, among them the notorious DDT, accumulate in the food chain and cause severe thinning of affected birds' eggshells. Their widespread use soon devastated the birds' ability to breed successfully, with Peregrines and Sparrowhawks particularly impacted. It's estimated that Peregrines plummeted to a low of just 385 breeding pairs in 1961.

Turning the tide

The second half of the twentieth century saw a change in the birds' fortunes. There was a definite shift in attitudes which led to the protection of birds of prey by law and a reduction in pesticide use. In 1954, the Protection of Birds Act was passed, giving legal protection to birds, their nests and eggs all year round, excepting game and food birds. Effectively this protected all birds of prey, apart from the Sparrowhawk, which had to wait until 1963. Meanwhile, a voluntary withdrawal of DDT and related pesticides began in 1962, with a total ban in place by 1982. Conservation and a series of reintroduction schemes also played an important role in helping to put raptor numbers on the road to recovery, with White-tailed Eagles an ongoing tale of conservation success.

Saving our Sea Eagles

White-tailed Eagles are also known as Sea Eagles as they often nest in trees along the coast, where they fish for prey. These are our largest birds of prey, with 2.5m wingspans, earning them the nickname of 'flying barn door'. However, habitat loss and persecution pushed White-tailed Eagles to extinction and the last UK-bred bird was shot in the Shetland Isles in 1918. The birds first began breeding again in Scotland as the result of a reintroduction programme, in which 82 young White-tailed Eagles were

translocated from Norway to the Scottish Isle of Rum between 1975 and 1985.

The first reintroduced White-tailed Eagles bred in 1983, and the first chick fledged in 1985. Further reintroductions to Wester Ross and the east coast of Scotland have secured the bird as a breeding species, with around 130 breeding pairs now in Scotland. In 2019, a five-year programme to introduce White-tailed Eagles to the Isle of Wight began when 25 birds were released in an effort to establish an English population. The first White-tailed Eagle chick hatched in England in 2023, more than 240 years since the country's last recorded chicks in 1780. It is hoped that the establishment of a population on Britain's south coast will also support populations in France, the Netherlands and Ireland.

↑ White-tailed Eagles often hunt along the coast.

Righting the wrongs

The change in the fortune of White-tailed Eagles is in many ways down to the dogged determination of people, not least the conservationist Roy Dennis, whose work was instrumental to the return of the birds to Scotland and who is similarly behind efforts to reintroduce the birds to England.

The decline and revival of Red Kites and Ospreys also tell a similar story, with both birds returning once humans intervened to protect and nurture them. Sadly some other species, such as the Hen Harrier, are still often the victims of illegal persecution and are not yet thriving as they should. But the conservation success stories of White-tailed Eagles and others demonstrate an important truth. While humans once may have caused near-extinctions, we also have the capacity to put things right.

Birds of the month

Red Kite

IN RECENT YEARS, Red Kites have made a remarkable recovery. Heavily persecuted throughout the nineteenth and early twentieth century, by the 1930s they were down to fewer than 10 breeding pairs, all clinging on in a remote part of Wales. This tiny population was carefully nurtured by a Kite Committee of conservationists and landowners. In 1989, there also began a series of reintroductions in Scotland and England, using young kites from Spain and Sweden. It proved incredibly successful and in 2015, Red Kites were moved onto the Green List as no longer being a bird of conservation concern. Red Kites can now be seen pretty much anywhere in the UK. Indeed, UK-born Red Kites have since been translocated to Spain to help bolster populations there.

TOP ID TIPS

Red Kites have an unmistakable forked tail that they twist and pivot as they fly. These are big birds, with wingspans of over 1.5m. Their size is most obvious when contrasted with 'mobbing' Carrion Crows, who jostle the kites mid-air in an effort to move them away from their territory.

The red in their name is for the rusty-red colouring, most evident on their bellies and upper tails. Seen from below, Red Kites have white patches on their wings. These great birds don't so much roar as mew, with a long call of *wheee* and often *whee-oo, oo, oo, oo, oo*.

WHEN AND WHERE TO SEE

All year round and in most of the UK, including urban areas.

Peregrine

PEREGRINES ARE THE world's fastest animal, reaching speeds of over 300kph in their famous stooping dive. With a little effort, you can see these record breakers right here in the UK. Traditionally Peregrines nest on cliffs and mountain ledges, but in recent years many have taken to our cities, nesting on tall buildings (see page 106).

The coast is also a good place to look for them, and here you might witness them playing chase with the local Ravens. Peregrines and Ravens seem to have a love–hate relationship, and both species show off their breathtaking flight skills in dramatic aerobatics. Juvenile Peregrines are also very playful with their nest-mates in the first weeks after fledging.

TOP ID TIPS

Peregrines are medium-sized raptors with intense-looking dark eyes, a strong hooked bill and bright yellow legs. They also have a thick, droopy dark 'moustache' on their white cheeks. Their upper wings and back are brownish-grey and they have a speckled breast and lightly barred belly. Juveniles are browner and scaly-looking, and their bellies have vertical streaks rather than horizontal bars.

Peregrines are noisiest during their courtship in early spring, and if you head to an urban nest site you may be able to hear their sharp squeaking. Once the chicks arrive, the screeching begging call of the chicks can alert you to the location of the nest.

WHEN AND WHERE TO SEE

Peregrines can be seen across the UK and in many cities all year round. For urban viewing sites, search online for 'Peregrine watchpoints'.

Sparrowhawk

THESE SMALL HAWKS are the raptors that you are most likely to encounter in your garden. If the general chatter and chirrup of birds suddenly stops, take a good look around as this often heralds the arrival of a Sparrowhawk. Usually, however, the hunt is swift and sudden, as a hunting Sparrowhawk often remains hidden away while eyeing up its prey, before swooping in without warning.

The Sparrowhawk's taste for garden birds has caused some people to worry that it is the pressure of this predation that has caused a decline in songbirds. But there is no evidence for this. As top predators, Sparrowhawks can only be sustained if there are enough smaller birds to feed on, creating a natural limit to the number of Sparrowhawks. Habitat loss, climate change and poor agricultural practice are far greater causes for concern.

TOP ID TIPS

Male and female Sparrowhawks are different in appearance. The female is a lot bigger, and has a brown back and wings, tinged with grey. She has fine brown bars on her pale belly, and on the undersides of her wings. The adult male, meanwhile, has a grey-blue back and upper wings, with rusty orange cheeks and belly bars. Juveniles of both sexes are much browner than adults, with thick brown barring on the belly.

Sparrowhawks are very much silent hunters but can sometimes be heard calling with a shrill *kew, kew, kew, kew*.

WHEN AND WHERE TO SEE

All year round in parks, gardens, woodlands and the wider countryside.

↑ Male Sparrowhawks have grey-blue backs and orangey chests, unlike the browner female (inset).

Kestrel

IF YOU SEE A BIRD of prey hovering, wings outstretched and with its long tail fanned wide, then you can be almost certain it's a Kestrel. These are super-sighted hunters with a remarkable ability to keep their heads still while hovering, even in strong winds. This enables them to pinpoint their prey before they swoop down to claim their prize. Kestrels are widespread across the UK and have a particular fondness for grassland habitats where they can best find a steady supply of voles, mice and small birds. These pretty little raptors were once our most common bird of prey, but have since been usurped by Buzzards. Numbers have declined in recent years and Kestrels are now on the Amber List of Birds of Conservation Concern, making them one to watch to avoid further declines.

TOP ID TIPS

Kestrels can be confused with other smaller birds of prey, particularly Sparrowhawks, but the distinctive hover often gives them away. Both males and females have red-brown backs, with this colouring continuing onto the upper wings, turning black as the wings narrow towards the tips. Rather than the barred bellies of Sparrowhawks, Kestrels have dark spots. Males can be recognised by their grey heads and tails.

Kestrels call with sharp squeaks, repeating *kee, kee, kee, kee, kee*.

WHEN AND WHERE TO SEE

Widespread in the countryside all year round and often seen from a car, hovering over hedgerows or fields.

Buzzard

BUZZARDS ARE BRITAIN'S most common raptor and these bulky birds of prey are often seen soaring high overhead on hot, sunny days. These are large birds, with wingspans up to 1.3m. You may see them hovering, but this is usually a brief affair and lacks the precision and grace of Kestrels.

Buzzards are also seen hunched and scruffy-looking, often described as 'hulking', as they perch on a post or fence or on the ground. They are opportunists, happy to hunt or scavenge for food. As well as catching rabbits, mice and other small mammals, they feed on carrion and are frequently seen feasting on roadkill.

TOP ID TIPS

Buzzards are large raptors, with broad wings. When seen from below they have round-ended tails, distinguishing them from Red Kites who have clearly forked tails. You may also be able to see their brown body, and this colour extends into their largely white underwings. The underside of their tail is also white, but the tips of the tail and around their wings is fringed with a dark brown that looks black from a distance. They vary in colour, with some individuals being almost white all over.

Like Red Kites, Buzzards have a mewing call that seems far too small and gentle for these great birds.

WHEN AND WHERE TO SEE

Can be seen all year round. Widespread, particularly in areas of open countryside, but can also be seen above urban areas.

August's challenge bird

Hobby

HOBBIES ARE SUMMER visitors that arrive in the UK in April and May, nesting and raising chicks until they return to Africa in September. They are supreme flyers with a stunning agility that enables them to catch fast-flying Swifts and Swallows, as well as dragonflies. These smallish raptors are similar in size to a Kestrel and their scythe-like wings can give them the appearance of large Swifts.

With dragonflies a particular favourite food, Hobbies are often found near water. And should the dragonflies be particularly plentiful, hungry Hobbies can gather in large groups of up to 50 birds. Watch for them skilfully catching their prey with their talons and eating it mid-flight.

Hobbies were once restricted to southern Britain, but their range is expanding ever further northwards and they now also breed in southern Scotland, giving even more of us a chance of seeing these fast winged-wonders closer to home.

TOP ID TIPS

Hobbies are quite elegant-looking birds, with long pointed wings and long-looking tails, which they often hold straight rather than fanned. Like Peregrines, they have a dark moustache-like colouring on their faces. Adult Hobbies have a reddish colouring around the base of their bodies, giving them the appearance of wearing red trousers when perched.

WHEN AND WHERE TO SEE

Hobbies nest in trees near open country, but are often seen hunting over wetlands, particularly old gravel pits. Can be seen from April to October.

Monthly musings

Rich pickings

While most birds have put breeding behind them by August, there are always those that break the rule. One interesting example is the Goldfinch. These pretty finches love small seeds, with Dandelion, Teasel and thistles all popular options. Thistles go to seed from late July onwards and, taking advantage of this, Goldfinches may hatch chicks as late as August to make the most of the harvest. Goldfinches usually breed twice in a season. The first clutch is typically laid in May, and with incubation through to fledging taking as little as just over three weeks, they are then able to nest again. In a good year, these busy birds can even nest for a third time.

What's up, me duck?

One bird that sometimes causes consternation in summer is the Mallard. With their yellow bills and emerald heads, male Mallards are easily identifiable, popular ducks at the local pond (see page 207). But come summer and it can seem like all the males have gone, with only the buff brown females to be seen. However, this is a case of mistaken identity. When ducks moult, they lose all their wing feathers at once rather than gradually like other birds. Unable to fly, they are vulnerable. So to give them a bit more camouflage, the first feathers grow back brown, making them look more like females. This is known as eclipse plumage. Once the flight feathers have regrown, the birds moult again and by late autumn they have regained their characteristic bright colours that identify them as male. Just in time for them to begin the process of courtship again!

↖ Goldfinches sometimes pick petals for their nests.

↓ A male Mallard in eclipse plumage.

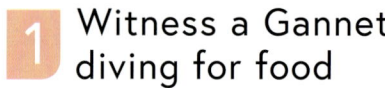

Top two must-dos

1 Witness a Gannet diving for food

Most people would rather not be called a 'gannet', given that it's slang for a glutton. However, this disparaging term really doesn't do justice to these supremely skilled and stunning seabirds. Gannets are our largest seabirds, with wingspans of up to 2m. They are effortlessly elegant with long wings, tail and neck, and long dagger-like bills. Their feathers are snow white with black wingtips and their heads are crowned in golden yellow in summer. But it's their diving that really sets them apart.

First, they fly high, circling until they spy their prey through the waves below. Then, from heights of up to 30m, Gannets will plunge towards the sea, reaching speeds of up to 100kph. Just before they hit the water, they fold their wings in tight, making their bodies as streamlined as a spear as they strike. It's perhaps the efficient precision of their fish-catching technique that has earned them the reputation of voracious eaters. But eating is of course a necessity and catching fish is crucial, particularly with hungry young to feed.

Great rocks for Gannets

Every year, from February, Gannets return to around 20 or so breeding sites around the coast of the British Isles. Just two of these are mainland cliffs, with the rest remote islands or offshore stacks. These breeding colonies or gannetries are internationally important, with more than half of the world's Gannet population breeding here. Bass Rock

↑ Gannets diving for fish off RSPB Bempton Cliffs.

→ Bass Rock is home to the UK's largest Gannet breeding colony.

off the east coast of Scotland hosts the UK's largest colony, the island becoming white with birds and guano (their poo) during the summer months. Further south in East Yorkshire, RSPB Bempton Cliffs nature reserve holds the record for the largest mainland gannet colony.

Although Gannets breed in just a few places, it is possible to see them almost anywhere on the coast and throughout the year. Between April and August it is easiest to see them near or on their gannetries, but Gannets will travel for food and it's worth keeping a look out whenever you can see the sea. During August and September when the young fledge and the birds migrate south, it can be a particularly good time to see Gannets diving. Look for them out at sea, flapping then gliding, often low over the water. Then watch for them flying higher and circling before they begin their rapid descent. The young Gannets are dark brown, but have the same distinctive shape and aerial aplomb as their parents.

2 Watch birds snap up a feast on Flying Ant Day

At some point in the long hot days of summer, there's a natural event when insects gather in such large numbers that clouds of them can be detected by weather radars. Commonly called Flying Ant Day, this is when swarms of new, wing-bearing male and queen ants emerge from their nests and take to the skies. It's a captivating event in its own right, but add to the mix a fleet of aerobatic birds making the most of this flying feast and it's a summer scene to be savoured.

Flying Ant Day usually occurs in July or August and follows a period of hot and humid weather. As weather will vary across the country, it's more of a Flying Ant Season than any one particular day. As might be expected, Flying Ant Days tend to happen earlier in the year in the warmer south and south-east of the UK.

Flights of fancy

The reason for all this activity is to enable the next generation of ants to mate and found new colonies. Ant colonies are headed up by queens who lay eggs, while worker females tend to this new generation. In summer, winged males and potential queens emerge and, in what's known as nuptial flights, the larger females will mate with several males. The queens then fly down to lay their eggs, nibble off their own wings, and begin a new colony of their own, while the males perish, their life's work complete. This mass event, with different colonies flying at the same time, maximises opportunities to mate and lowers the chance of inbreeding.

All this activity is like manna from heaven for hungry aerial insect-eaters such as House Martins and Swallows, who can be watched hunting in the huge clouds of ants. If the event happens before their return to Africa, Swifts too will join the throng, all skilfully swooping and scooping up ants with ease.

Party time

Gulls, meanwhile, have become notorious for their taste for ants, with some commentators suggesting they get drunk and giddy, as a result of formic acid found in the ants. While the science here is shaky – the acid concentrations are too small to have an effect – anecdotally it's clear that this abundance of fast food is a cause for excitement. Two of our most commonly seen gulls, Herring and Black-headed Gulls, will flock to food, calling raucously as they set about feasting, and it's this revelling behaviour that seems to have fuelled the myth. But given a sudden surfeit of tasty treats, who wouldn't indulge in a bit of revelling, whether human or bird?

↑ In summer, ants grow wings for nuptial flights.

How to help

Farming for people and wildlife

The RSPB estimates that there are around 38 million fewer birds in UK skies than there were 50 years ago. This decline is a shocking statistic but it's also reflected in the ever-increasing number of birds joining the UK's Red List of birds that need active conservation to help them. While 50 years ago people may have expected to hear Cuckoos or listen to Nightingales with ease, we have to try a lot harder to do so these days.

Down on the farm

Farmland species have been particularly hard hit, and it's estimated that around half of the birds, insects, plants, amphibians and reptiles found on farmland have vanished since the 1970s. The decline in farmland birds is horribly stark, with species such as Turtle Doves dropping by 99 per cent between 1967 and 2020. Turtle Doves, together with Grey Partridges, Tree Sparrows, Yellowhammers, Linnets and Skylarks, were once fairly common birds, but all have suffered huge losses and are now on the Red List.

The main drivers for these declines are changes in farming practice, with intensive agriculture, together with an associated increase in pesticides and fertiliser use, proving catastrophic. The decline in farmland species isn't just bad for nature, it's also bad for us. A healthy functioning ecosystem should be self-sustaining with clean water, clean air and fertile soil. But intensive farming involves a downward spiral of harmful pesticides that wipe out pollinating insects and other wildlife, and a reliance on expensive fertilisers that can leach into our rivers, streams and other waterways, causing pollution.

↓ Turtle Doves have suffered dramatic declines in the UK.

Positive action for birds and people

In the UK, around 70 per cent of land is used for farming. This presents a huge opportunity to effect positive change. And, of course, many farmers are already working hard to promote biodiversity and protect wildlife, working independently or with organisations such as the Nature Friendly Farming Network, the Organic Farmers Association and the Soil Association.

The RSPB's Fair to Nature mark is a certification scheme for food and drink. Products which have the certification are committed to sourcing their ingredients from farms that meet a rigorous set of criteria, including managing 10% of their land for nature, by providing good-quality wildlife habitat. Hope Farm in Cambridgeshire is an RSPB Fair to Nature farm. It's run by the RSPB and demonstrates how it's possible to manage a profitable farm that also makes space for nature. Since the RSPB took over the farm in 2000, birds including Yellowhammers, Skylarks and Linnets have at least tripled in number.

One way to help bring back farmland wildlife is by supporting those who farm with nature in mind, so look out for foods that are certified as helping nature and buy organic wherever possible. There is also a need for greater government support for farmers to reduce pesticide use and to adopt more nature-friendly practices. So if you've got a spare five minutes, why not write to your local elected representative to urge them to take action?

↑ Grey Partridges are on the UK's Red List.

Myth of the month

Something in the air

There is an old rhyme that states, 'Hawks flying high, means a clear sky. When they fly low, prepare for a blow.' It is true that we often see large raptors such as Red Kites and Buzzards flying high on a hot summer's day as they soar upwards on the thermals. Conversely, when bad weather approaches, there's a corresponding drop in air pressure, making the air less dense, and this does make it more difficult for birds to fly high. There's a similar saying about geese: 'When the goose flies high, fair weather. If the goose flies low, foul weather.' Again, it's likely the geese are taking advantage of easier flying conditions when they fly high.

But does a low-flying bird really mean that bad weather is on its way? There is evidence that birds respond to changes in atmospheric pressure. Research in North America, for example, found that migratory songbirds were more likely to depart when atmospheric pressure had been rising over the past 24 hours, when the weather is more likely to be fair. It's also likely that birds will fly to shelter to avoid storms or heavy rain, meaning they could be seen 'flying low' as they head to somewhere safe. So there is certainly a ring of truth to these sayings, although of course any bird will be 'flying low' regardless of conditions if it has just taken off, or is coming in to land!

↑ Red Kite flying high.

SEPTEMBER | 9

Journeys long and short

September is a big travelling month for birds, with significant numbers of species, and millions of individuals, shifting away from the breeding areas on journeys long and short. Every day can turn up something different and interesting, wherever you are, and you can revel in the sheer variety.

BIRDWATCHERS LOVE September because it is a time of moving and mixing. We can enjoy a sumptuous mix of summer visitors and winter visitors; the summer visitors predominate at the beginning of the month, and the winter visitors become more frequent towards the end. Some birds are September specialists, hard to see at other times of year.

When people see Swallows gathering on wires in September, ready for their great southward journey, there is a tendency to feel that our birdlife, shorn of these lively companions and others such as martins and warblers, is about to be depleted. But that is an entirely false impression. It is merely a changing of the guard, with different faces. Masses of wonderful birds that breed further north in Eurasia see the British Isles as their preferred winter destination. We are in the fortunate geographical position of being on the western side of the continent, where winter frosts are comparatively mild, snow is moderate, and it is decidedly damp. That is a big attraction.

Many of the birds leaving us are transcontinental migrants, such as Swallows, flycatchers and some warblers, on their way to winter south of the

→ Whooper Swans start to arrive in the UK in September after breeding in Iceland.

← An autumn flock of waders, with Black-tailed Godwit, Redshank, Turnstone and Knot.

Sahara. On the other hand, many of the new arrivals are from closer at hand; for example, northern Europe. Their journeys are shorter, though no less dramatic. Whooper Swans, which appear at the month's end, may migrate from Iceland in a single day and expend little effort in the process. The crossing from Sweden to the UK is quite an undertaking for a Goldcrest, though – these 5.5g birds are not exactly built for long-distance flight.

Mud, glorious mud

One of the best places to witness the swollen ranks of incomers is on a muddy estuary. In summer, these are quiet for birding, being better for human holidaymakers and day-trippers, with the odd desultory Redshank, Shelduck or Black-headed Gull lurking around in the sunshine. But as autumn kicks in,

a cornucopia of wonderful gulls, wildfowl and, especially, waders descend on our saltmarshes and tidal estuaries, making them thrilling and lively places. Numbers increase through the autumn and into the winter. Some estuaries support many thousands of birds.

You might ask: 'What's the attraction of mud?' The answer is that it provides an enormous amount of food that's reasonably easy to reach. Estuaries are attractive because they form the point at which a river, carrying all its nutrients, discharges into the sea. Plus, the action of the tide tends to rinse and sift even more nutrients through the mud, meaning that it is a very productive ecosystem. A square metre of mud may support more than 4,000 small organisms, which are mainly of three types: worms, crustaceans and molluscs. The crustaceans include shrimps and small crabs, while the molluscs encompass burrowing clams,

→ Grey Plovers have short, stout bills and catch their prey by sight.

↙ The Curlew's curved bill is ideal for reaching into the burrows of crabs.

cockles, mud snails and other shellfish found in our intertidal zone.

If you consider the productive mud, and you add in all the usual furnishings of an estuary, such as creeks, sandbanks, areas of shingle and seaweed, you get a varied habitat which can support many different species of birds. If you get the time to watch an estuary, you can see all the different ways the birds use it.

Shaped for feeding

With so many species present on the mud, there has to be some variety in the way in which they obtain food, or the competition would be intolerable. Waders have evolved different body forms, especially bills, to ensure that each has specialisations, and they also practise their own techniques.

One obvious difference is in size. A Curlew towers over a Dunlin, and a Redshank is somewhere in the middle. A Black-tailed Godwit is a proper wader,

going into belly-deep water and probing its long bill into submerged mud, whereas a Dunlin can barely paddle. A Curlew has a long, downcurved bill, which helps the bill tip's turning circle, enabling the bird to reach through and behind barriers. Greenshanks and Spotted Redshanks scamper and swim in the shallows for fish. Knots use their fairly long but quite thick bills to 'plough' through the mud, hoping to touch prey. Avocets, where present, scythe their upswept bills through deep water, hoping for the same result.

One of the most interesting distinctions is between sight feeding and touch feeding. Plovers are the main exponents of sight feeding. They have large eyes, and survey the mud from a still position, watching for the odd crab moving about, or perhaps the telltale movement of a worm excreting. Once they detect something they run towards it and grab it. The Grey Plover, a large species, can crush and eat larger items than the tiny

151

Ringed Plover. While many other waders do feed by sight, they can also use the tips of their bills to detect food by touch (see page 159). Birds using this ability can feed in tight flocks when together as they won't disturb each other; plovers can't do this and may even have their own territories on the mud.

Passing through

Another delightful feature of September, the moving month, is that a wide variety of species pass through that we probably won't see at other times of year. These include passage migrants, birds that don't breed or winter here, but are merely transients. Among those passing through are popular birds such as Black Terns, Little Gulls, Wrynecks and Red-backed Shrikes, and the chance of finding one of these special visitors adds a little spice to the birdwatching day.

New beginnings

In September, you might not expect birds to be thinking about breeding, but several species have their main pairing-up month now. Two are very familiar. House Sparrows (and Tree Sparrows) often spend a few weeks away from their colonies in August, but when they return young birds set up a territory by a nest hole and solicit for a mate. Young Rooks also form pair bonds, attending the colony and flying together with a prospective mate. They need to be ready for an early start to the breeding season, in February or March.

After a summer lull, song begins again in September, with birds aiming to form and protect winter territories. Robins begin, as do Dunnocks and Wrens, as well as some more unusual birds such as Cetti's Warblers. Great Tits occasionally sing briefly, but nothing too serious yet.

↑ September is a wonderful month for seeing rare birds such as this Wryneck.

Birds of the month

Knot

IN THE UK, the Knot is famous above all for the flocks it forms in some large estuaries, which swirl in enormous numbers near their roost sites. When they all turn, their undersides may catch the light and glint; these aerial spectacles are one of the great sights of birding (see page 160). Knots are widespread on estuaries in winter, but impressive numbers gather only at certain big estuaries, such as The Wash on the east coast of England, and Morecambe Bay between Lancashire and Cumbria. This species is a touch-feeder which inserts its bill into soft mud in order to detect bivalve molluscs. In the summer, our birds go a long way to breed, often to Greenland or Arctic Canada, where they mix in with Polar Bears. That must make a damp British estuary seem dull.

TOP ID TIPS

It sports a stunning orange-red colour in the breeding season, but we hardly ever see that, because it moults into a muddy grey-brown in winter. Tricky to identify, it is larger and fatter than a Dunlin with a distinct pale eyebrow, and it lacks the whitish spots of the similar-sized Grey Plover, often seen nearby. The legs are greenish. It's an undemonstrative bird as an individual, with no loud ringing call, just a flat *knut*. It characteristically leans down when feeding.

WHEN AND WHERE TO SEE

Only present in winter on muddy estuaries, mainly larger ones. Migrates to the Arctic in summer.

← A juvenile Knot.

Lapwing

OUR ONLY WADER with a wispy crest, the Lapwing is an unusual species, also being distinct for its strikingly rounded wings and strong attachment to agricultural fields as opposed to muddy estuaries in winter. In spring, it dazzles with its gorgeous iridescent plumage and a truly wonderful display-flight, in which it tumbles over its territory with crazy rolls and twists, following an undulating path like a rollercoaster, and giving its sweet, pleading call. In the winter it is found in large flocks on pastures and fields, where it seeks worms by sight, using a stop-and-run method typical of a plover. For breeding it makes no more than a 'scrape' on the ground for its nest and, as is typical for most waders, it lays four eggs.

TOP ID TIPS

Easy to identify as nothing looks quite like it. The iridescent green and purple on the back shimmers in the light; also take note of the lovely caramel-coloured patch under the tail. The crest is longest in the male; juveniles have the shortest crests. Flies with a very noticeably floppy action and at a distance can look like ash blowing in the wind. Call is a wheezing *pee-wit*, with a slightly complaining air, but the sounds can turn whooping and ecstatic during display.

WHEN AND WHERE TO SEE

Common and widespread. Found in farmland areas all year, as well as marshes and bogs. In winter in fields and grassland.

Oystercatcher

THERE'S NO SUCH thing as a softly spoken Oystercatcher. It has two types of call: loud and very loud. There's no such thing as a hidden Oystercatcher, either, with its unmistakable black-and-white plumage and long, straight orange bill which makes you think it is carrying a carrot. Many birdwatchers are grateful for at least one wader that's easy to identify, big and bold, a common sight in the mix of waders on winter estuaries. It is curious for having two distinct methods of feeding, hammering and stabbing. Hammerers stride up to shellfish and simply beat them open, aiming at weak spots and breaking the shell or hinge. Stabbers attempt to creep up on their prey and try to stab at and slit the adductor muscle that holds the bivalve shut. Individuals tend towards one feeding method or the other, and bill shape varies slightly in accordance with this. Oystercatchers are also interesting because they bring food to their newly hatched young, which take some time to learn these feeding techniques.

TOP ID TIPS

Unmistakable. The call is a shouted *ke-BEEK*, uttered as though there was an emergency.

WHEN AND WHERE TO SEE

A common wader throughout the UK, found both on the coast and inland, on fields and meadows, becoming commoner as a breeding bird as you go north. Prefers to nest on patches of shingle, where it can hide its eggs. Many Oystercatchers from further north winter in Britain and Ireland.

Curlew

IN COMMON WITH many waders, the Curlew swaps an upland breeding habitat in summer for an intertidal or beach existence in winter. In the uplands, it breeds on bogs and wet moors, where its gorgeous, ecstatic bubbling trill is a feature of the soundscape. It performs a display in which the wings are beaten fast and stiffly, before it slowly glides downwards. In the winter, it uses its long bill to probe and investigate mud and weeds, and it often eats crabs. One study found that females, who have longer and more curved bills than males, ate more shellfish deep in the mud, while the shorter-billed males caught more crabs. Curlews, especially males, also feed on pasture adjacent to estuaries, where worms are often on the menu.

↑ Unlike the Curlew, the Whimbrel has dark stripes through and above the eye.

TOP ID TIPS

A huge wader that towers over its colleagues at roost, and the long, downcurved bill removes any doubt. The only identification issue is with the very similar, slightly smaller Whimbrel (see photo above), which has a straighter bill with a kinked tip, and a brown stripe through the eye with a contrasting paler 'eyebrow'. The call of Curlew is a ringing *cour-lee!*, the origin of the name.

WHEN AND WHERE TO SEE

A common and widespread wader in winter, occurring pretty much all around our coasts, including on beaches. Often numerous on estuaries. Breeds inland on bogs and uplands, especially in the north, but numbers are declining.

Dunlin

ON THE WINTER ESTUARY, this is the overlooked wader, often present on the mud in large numbers. It is very small and scampers around in tight-knit flocks, whose whirring activity looks exhausting, and makes the birds look a bit like clockwork toys. It very easily takes flight, spooked by tide, disturbance or imagination, and flocks fly quickly and in agile fashion, highly coordinated, with quick turns. It seems to pick from the mud with a rapid sewing-machine action, but is capable of probing deeper for worms and bivalve molluscs. Although most familiar as an estuarine species, the Dunlin does breed in Britain, mostly on upland moors where it has a song that resembles several blasts on a referee's whistle.

↑ In summer plumage, the Dunlin seems to have dipped its belly in black ink.

TOP ID TIPS

In the depths of winter, a rather colourless wader, greyish-brown above and white below, notable for its small size, fairly long and slightly curved bill, and black legs. But in spring, summer and autumn it bears a black belly patch, so completely dark that the bird looks as though it has bathed in oil; it also becomes quite attractively reddish-brown on the upperparts. The call is a soft, short, rolling trill.

WHEN AND WHERE TO SEE

In winter, often abundant on mudflats and estuarine saltmarshes, especially on the coast but sometimes in shallow pools inland. In the breeding season, it is mainly found on upland bogs.

September's challenge bird

Sanderling

HOW WOULD YOU fancy a life spent running along the beach all day, flirting with the breaking waves when it's stormy, and drinking in the wide-open spaces and wildness? If that sounds perfect for you, then you have cause to envy the Sanderling. This wader is not drawn to the melee of the winter estuary, crowding in with Dunlins and Knots. The Sanderling instead is a bird of sandy beaches, a separate niche. One of its main feeding methods is to forage in the backwash, so it is constantly dodging the waves, taking items from the moist sand left behind. To do this it needs to be alert and very quick, and you almost always see flocks of these birds running fast, their legs ablur. Unusually, they have entirely lost their hind toe in the pursuit of running speed, and have just the three toes facing forwards. Their main foods are bivalve molluscs and crustaceans, such as sandhoppers, which they obtain by picking and probing (see opposite).

TOP ID TIPS

Its brilliant washing-powder white plumage on the underside really stands out. It has a black bill and legs, and when it flies it shows a more noticeable white wing-bar than a Dunlin. The typical running-about-on-beaches habit is a giveaway. The call is a soft and unremarkable *wick*.

WHEN AND WHERE TO SEE

Widespread on sandy beaches over much of the UK. It breeds in the high Arctic, so it's a winter visitor, September to April.

Monthly musings

The Sanderling's touch

The fast-moving, quickly probing Sanderling (see opposite), dodging the waves as it feeds, has some remarkable foraging abilities. Research has shown that it uses a quick taste analysis to identify areas with food, based on secretions from worms and molluscs. It also has such a sensitive bill tip that it can detect the movement of prey in the sand up to 2cm away. As the bill tip is inserted, it displaces water in waves away from it. If something is in the sand, those waves reflect back and show the bird there is food nearby – a kind of remote sensing.

↓ The Green Woodpecker is one of Britain's most specialist feeders.

The Green Woodpecker's diet

Not many birds are entirely specialised in their overall diet, but the Green Woodpecker is a familiar species that is. We almost always see it on the ground, perhaps on pastureland, sometimes even on a garden lawn, where it may spend many minutes in the same place. Its food fad is ants, and in places ants can comprise almost 100 per cent of its diet. There are 50 species of ants in Britain, so perhaps there is some variety in flavour! In winter, Green Woodpeckers still eat ants which are active in the soil and have been seen digging down to 85cm through snow to reach them. One woodpecker family fed 1.5 million ants to their growing family during the nestling stage.

Top two must-dos

1 Watch a whirling wader spectacular

Estuaries are rich and abundant places for waders to feed, but they do have disadvantages – well in fact, two per 24-hour period: high tides. Although the height of the high tide follows the phases of the moon, from springs (highest) to neaps (lowest), the more seawater over the mud, the smaller the feeding area available to waders. The higher the tides are each day, the more inconvenient they are, and the less time they permit a wader to feed.

But what do waders do when the tide covers their food? The short answer is that they find their way to anywhere nearby that remains dry at high tide and is safe to settle down on, such as an island, a jetty or exposed rocks. Some go to inland pools and fields. Here they roost and while away the time. However, prior to that, they do something spectacular, similar to what Starling murmurations do before they finally settle down at night. To a birdwatcher, this makes for compelling viewing.

Wader roosts are more complicated than Starling roosts, because there are more species involved than just one. A Dunlin, with its small size and short legs, tends to go to roost before a Curlew, which can still utilise the shallows. Estuaries are often big and varied places where the water may come in at different rates, meaning that a bird in one part may come to roost at a later time than another of the same species. This means that a roost is a place of constant rearranging and jostling, with birds becoming cramped and flustered.

Partly for this reason, and partly, perhaps, to advertise the presence of the roost, waders often swirl about for quite some time before settling, and to a lesser extent afterwards. When large numbers are involved, this can be a truly remarkable sight. At Snettisham in Norfolk, for example, 100,000 waders, mainly Knots, perform these aerial manoeuvres over the moving water, and something similar happens on Morecambe Bay, Lancashire. However, the same phenomenon can be seen on most

↓ Knots at Snettisham in North Norfolk performing their famous pre-roosting aerobatics.

estuaries, and most wader species are involved, although only Dunlins, Knots and Grey Plovers are the real swirlers.

The surefire way to enjoy the spectacle is to get your timing right. Select a daytime spring tide and arrive about an hour before high water. If it's also a still day, you can enjoy the best of the action, and you'll get the bonus of the fabulous whistling calls of these fast-moving birds.

2 Look for Swallows on wires

If you want to know that it's autumn, look on the overhead wires. Not far from you, somewhere, there will be Swallows gathering in migratory parties, many of which rest on high perches during the day. It is a sign of the passing season; one Swallow doesn't make a summer (see page 57), but a party of them does make the autumn. If you take a close look, you might notice some of the birds have longer tail streamers than others; adults have a longer tail than juveniles.

Swallows are famous for their migration, and rightly so; it is one of the longest for a small bird in the world, taking our birds from the UK down to South Africa, near the Cape region where, oddly, they mix with Russian ones. It's a journey of nearly 10,000km. The migrants usually fly by day and are happy to pick up random flying insect meals along the way, as befits a marathon. Only when they overfly the Sahara Desert do they fly at night, partly out of the need for speed, to reach more productive feeding grounds and refuel, and partly to keep cool.

Now is the time for the birds to head south, starting in the first few days of the month. At first, the Swallows' daily journeys are very short, just a few kilometres, and they will roost not far from where they spent the previous night. By the month's end most individuals will have left the UK, but they aren't really in a hurry. They cross the English Channel and travel down the west of France, until they meet the Pyrenees. Many then cross to the Mediterranean and drop down the east coast of Spain. They reach North Africa in early October and tropical West Africa later that same month. It is often not until December that they reach their winter quarters. Three months later, they will be on the way north again.

In common with every other migrant, their northbound journey will be much faster than the southern one, as they are in a hurry to find a territory and begin their breeding season, and it has been estimated that northbound Swallows travel about 320km per day.

Recent research using geolocators has revealed something quite unexpected: that, at least sometimes, pairs migrate together. This is thought to be very unusual in small migrant birds. But it's a thought to lift the spirits. On those wires there could be adult pairs about to embark on a far distant journey in each other's company.

↑ Swallows of all ages (and House Martins) gather on wires ready to migrate.

How to help

Make a home for nature

However keen you are on gardening for wildlife, you can always do something extra to help nature on your doorstep. As wildlife gardening has taken off over recent years, so people have come up with new ideas, many of which are great fun to do. Here are just a few. The RSPB has a wealth of gardening activities on its website to help birds, insects and other wildlife.

Insects

Whatever sort of garden you have, or even if you just have a windowsill, you can help insects. In their mini world, the smallest adjustments make an impact, and of course more insects means more food for our insect-eating birds.

Have you ever heard of a hoverfly lagoon? Hoverflies are fantastic pollinators as well as cool insects that can beat their wings 80 times a second, but many of their grub-like larvae develop in pools of stagnant water, including one species with an amazing name, the Batman Hoverfly. A hoverfly lagoon is just a small container of water, plus decaying leaves and other organic matter, and can be left in a quiet corner.

There is a current fashion for bee hotels and bug hotels, and they are all brilliant. A few holes drilled into wood can make a good hotel for solitary bees. You can make a special hotel for multifarious bugs, looking a little like a bookshelf, with

↓ Solitary bees, such as Red Mason Bees, are easily encouraged to breed in the garden.

all sorts of bug-friendly things packed in the 'shelves', such as logs, stones, straw, sticks, dry leaves, bark and cardboard (for lacewings). You can also make a log pile, which will be a boost for beetle larvae and much more. You might see a Wren exploring your bug hotels and log piles in midwinter, and extracting any guests that are not tucked away securely enough.

Home to roost

Nest boxes aren't just for nesting in – they can also offer a cosy place for birds to sleep as the nights become colder. A hole-fronted nest box may attract a lone Blue Tit or Great Tit, or perhaps even a communal roost of Wrens. A male Wren will call to attract others to a suitable roosting place in his territory – more bodies packed in together will generate more life-saving warmth, and in extreme cases one nest box could shelter dozens of Wrens. However, the territory-holder prefers his guests to be female. Rival males might be refused admittance, especially on milder nights.

As well as nest boxes, you can also buy or make special 'roosting pouches' which are designed to be sleeping places only. Place them in particularly well-sheltered spots and keep an eye on them at dusk to see if you have any visitors.

← Roosting pouches are a comparatively new way of helping garden birds through the winter.

Myth of the month

Gulls coming inland

Everybody other than birdwatchers calls gulls 'seagulls' and they probably always will. You can tell somebody until you are blue in the face that lots of gulls occur inland, and they will still call them seagulls. There is also a widespread saying along the lines of 'When there is a storm at sea, the seagulls come inland.' Where does all this come from?

One thing we can be sure of, and that's the fact that gulls are extensively found inland, especially in the winter. You'll see Black-headed Gulls everywhere, from parks to arable fields and from rivers and lakes, and the Common Gull often joins them on playing fields. Herring Gulls and Lesser Black-backed Gulls now nest in many inland cities, including London and Bristol. They are also drawn inland to rubbish tips and landfill sites, and often roost on large reservoirs a long way from the sea.

So, is there any truth to the 'Seagulls and Storms' legend? Perhaps surprisingly, there is. Inland gulls are a relatively recent phenomenon. The first awareness of Black-headed Gulls evacuating the sea dates back to the 1880s, when a succession of hard winters forced gulls into central London, and into people's consciousness. It was more of a freeze than a blustery storm, but there was a connection.

Of course, gulls are principally seabirds, and they are largely immune to storms, riding them out expertly, so it would take something apocalyptic to shift them too far inland, just as it would take something apocalyptic to expunge the word 'seagull' from everyday use.

↓ Black-headed Gulls cannot resist the lure of worms exposed by a plough.

10

OCTOBER

Hoarders and feasters

Autumn's bountiful harvest of berries, nuts and seeds, abundant in our trees and hedgerows, means rich pickings for birds. It's a time of gathering and gorging, which makes for plenty of birdwatching opportunities for us!

COME OCTOBER and autumn is definitely upon us. Our trees are swathed in golden hues, their vibrant red and yellow leaves getting ready to tumble as the wind blows wilder. Berries abound, from the dark-blue sloes amid spiky Blackthorns to the lively oranges of Rowan and Dog Rose. There's a definite chill in the air and we may even feel the first frosts. The future will be colder, and nature acts accordingly. It's time to reap the benefits of the harvest and prepare for the tough winter ahead.

← Waxwings are sporadic winter visitors to the UK.

↓ Jays are beautiful members of the crow family.

Are you winter-ready?

First up are the hoarders. Jays are flamboyant members of the crow family, with jewel-like blue flashes on their wings. Far shyer than their all-black relatives, they are drawn into our gardens in autumn. The Jay is an eye-catching visitor with a discordant call, and apt to send feeder regulars flying as it sweeps in. These are birds on a mission, determined to find acorns wherever they may be, scouting parks and gardens for as many as they can find. From late September to early November, Jays collect acorns then hide them away to eat at a later date.

Hiding places are many and varied, dug in the ground and buried under leaves, jammed between tree bark or hidden in holes in trees.

Smaller birds too will cache food to eat later. In the garden, keep an eye out for diminutive Coal Tits. Wary of bigger birds, they will steal away seeds from the feeders to eat at a more leisurely pace later. Another hoarder is the Nuthatch. These pretty tree dwellers hide food for winter, wedging acorns, hazelnuts and seeds into bark fissures for safekeeping. Unlike Jays, who hide their acorns far and wide, Nuthatches keep them within a smaller area. Highly protective of these stored supplies, they will feistily defend their patch, remaining territorial over autumn and winter.

↓ An abundance of berries is great for birds.

→ Waxwings feast together in flocks.

Harvest time

A second strategy is to feast on the autumn harvest. The first berries start fruiting from late August, with blackberries and Rowan berries two early examples. By October, this swells to a banquet with Hawthorn, Blackthorn and more all bearing fruit, soon to be followed by Ivy and Mistletoe. These berries are packed with energy that helps birds through the winter. Birds will eat the easier to digest, 'short shelf-life' berries such as blackberries first. Those that stay fresher longer, or with a mild toxicity such as Ivy, are left until there's less choice. Whether conscious or not, it's a clever strategy, ensuring the berry supply lasts as long as possible into the winter.

The sudden abundance naturally brings in a variety of birds. It's not hard to see how the brightly coloured red, orange and yellow berries act like neon advertising hoardings, alerting birds to the harvest as they fly overhead. It's thought that black berries such as Juniper and Ivy may have an ultraviolet reflectance that is visible to some birds, although not to us.

A berry good plan

But of course berry bushes aren't here solely to feed birds. The berries contain the seeds for the next generation and the birds are a means of effective dispersal. While most seeds are scattered after passing through a bird's gut, one plant relies on its berries' sheer stickiness. Mistletoe is a parasitic plant that grows in the branches of

trees. Its sticky seeds fix to the bills of the birds and are 'planted' when the birds wipe their bills clean on a branch.

It's not just our resident birds that reap the harvest, with a wave of migration seeing newcomers fly in from colder countries in northern Europe. Here come the hordes of hungry birds on the hunt for food. Fieldfares and Redwings are our 'winter thrushes', which flock to the UK from Scandinavia in autumn and feast on berries during their stay. Another bird that may come for the berries in winter is the punky-looking Waxwing. These beautiful birds turn up only when the berries in their northern homelands of Russia and Scandinavia run dry. In such a 'Waxwing winter' we can enjoy sporadic flocks popping up, gorging themselves until the berries run out. Autumn also sees the arrival of large numbers of familiar species including Blackbirds,

Song Thrushes, Chaffinches and Robins. These are birds that breed in northern and colder parts of Europe but fly here for our relatively milder winter, joining our resident populations. Birds know no borders and are motivated by the need to find adequate food and shelter wherever it may be. The number of these so-called partial migrants varies year on year depending on the weather. So come the colder months, it may well be that 'your' regular garden Robin is not a long-term resident but a visitor from overseas.

Noises in the night

October ends with Halloween and, although it's unlikely you'll see these birds flying with witches, it is a good time of the year to listen and look out for owls.

Tawny Owls are at their noisiest at this time of year. From early autumn,

these birds set out about claiming their territories, calling out at night with a suitably spooky hooting *hu-hu-hoooooo*. Young Tawny Owls are moving away from their parents and looking for territories of their own. And as the young owls branch out to stake their claims, the older birds call to defend their patch. Because of their nocturnal nature, you are more likely to hear a Tawny Owl than see one.

If you do manage to see one you might be struck by how quietly it flies. Like other owls, a Tawny Owl's wing feathers are specially adapted to minimise the amount of noise generated as it flies. Owls also have large wings relative to their body size, which means they don't need to flap their wings as much, enabling them to glide silently by. These adaptations help owls hunt at night, moving under the cover of darkness to stealthily catch their unsuspecting prey.

Barn Owls too are silent flyers, and they could also be said to have a cartoonishly scary call. Their other-worldly shriek can alert you to their presence and aptly Barn Owls are also known as screech owls. Barn Owls can be seen year-round at dawn and dusk hunting for food, quartering fields. However, as the days get shorter and the weather gets colder, Barn Owls will hunt in the day too, making them potentially easier to see at this time of year.

↑ Barn Owls hunt for voles, mice and other small animals.

Birds of the month

Jay

A JAY'S SCIENTIFIC name is *Garrulus glandarius*, which very appropriately is Latin for 'chatterer of the acorns'. This bird's love for acorns is legendary. You might see one flying purposefully overhead, carrying a couple of acorns in its bill and a few more in its throat pouch. Jays shriek repeatedly when disturbed. It's a harsh, grating call that doesn't sit well with the bird's dashing good looks!

Although common in most parts of the UK, Jays are only seen intermittently. These are shy birds, largely unnoticed until autumn, when they suddenly swoop into our parks and gardens gathering acorns. Their fleeting appearances and flamboyant colouring can make Jays seem like exotic visitors, with every sighting a special treat.

TOP ID TIPS

An angry-sounding raspy screech can be the first thing to alert you to a Jay. Despite their pretty colouring, Jays can be surprisingly hard to see. They have pinky-brown feathers, with azure-blue wing patches and black tails. In flight, look out for their rounded wings and a distinctive white patch on their rump. A close-up view of the head, with its drooping black moustache, can be especially rewarding. Jays can also raise their crown feathers into an impressive crest, which they do when alarmed or displaying to another Jay.

WHEN AND WHERE TO SEE

All year round in woodlands, but also in parks and gardens, especially in autumn.

Nuthatch

THIS IS THE UK's only bird that can go down a tree trunk headfirst. It's a nifty trick, made possible by strong legs and claws, and is fitting for a bird that spends much of its time working its way up and down trees, ferreting out food with its sharp, pointy bill. Nuthatches eat insects, nuts and seeds. In winter, they will also visit feeders in gardens.

Like Jays, they store food for winter, often wedging seeds and nuts in the bark of trees. The name 'Nuthatch' is actually related to the word 'hack' and refers to their habit of striking or hacking away at the wedged nuts and seeds as they battle to get to the edible insides.

TOP ID TIPS
Nuthatches are beautifully robust sparrow-sized birds that are orangey underneath and grey-blue on top.

Their faces are catwalk-ready, with a dramatically bold dark eye-stripe running from their dagger-like bills to the back of their heads. Males and females are similar, although males have richer, redder feathers at the base of their bodies. Juveniles are a little drabber in colour.

Nuthatches can often be heard before they are seen and they call frequently with a loud, repetitive *zit, zit, zit, zit*, as well as a longer, prettier-sounding whistle, sounding a little like *tweet, tweet, tweet*.

WHEN AND WHERE TO SEE
All year round in woodlands, parks and gardens.

Treecreeper

TREECREEPERS ARE mouse-like little birds that can be seen scurrying up trees, pecking away and holding their longish tails stiffly as they balance. Typically a Treecreeper starts at the bottom of the tree and works its way up and around the trunk, so once sighted, you can follow its progress until it flies down and starts the process again with the next tree along. It's an endearing habit that has coined the collective term for a group of Treecreepers, a 'spiral'.

Treecreepers are quite common across the UK but, being well-disguised woodland dwellers, they are often overlooked. Look out for them in mature woodland where they are most likely to be found indulging their taste for bark-dwelling invertebrates.

TOP ID TIPS

Treecreepers are well camouflaged with mottled brown feathers on top and a snowy white underside. Like Nuthatches, they have sharp bills to help them probe the bark for grubs, but a Treecreeper's is thin and curved. They have a noticeable eye-stripe, but this is far less flashy than a Nuthatch's black stripe, being a soft white brow above their eyes. Males and females look alike, as do juveniles.

Their call is a high-pitched *sree, sree, sree* often described as having a buzzing quality to it. Their song is similarly shrill, but can sound like a Blue Tit's song with a trill at the end.

WHEN AND WHERE TO SEE

All year round wherever there are mature trees.

Tawny Owl

BRITAIN'S COMMONEST owl, the Tawny Owl, is found in woodlands, parklands and even gardens with sufficiently mature trees. Being nocturnal, it roosts hidden in trees during the day, coming out to hunt at night.

Tawnies are patient hunters, watching from a perch until a mouse or vole passes, then swooping silently to take their prey. But that's not all they will eat. They can take fish from ponds, prey on roosting birds and will also eat worms when wet weather makes them easy to find. Their dispassionate efficiency does not make them popular with other birds. If a Tawny Owl is discovered by small birds during the day, it will be mobbed without mercy until it flies away.

TOP ID TIPS

With its big, round head and long, plump body, the shape of an owl is unmistakable. As the name suggests, Tawny Owls are a dappled blend of brown tones, darker on the back, wings and face and lighter on the belly. Like all owls, apart from the Snowy Owl of the far north, males and females have similar markings.

The *twit-to-whoo* associated with Tawny Owls is actually two separate calls. The male calls with a long fluting hoot, *hu-hu-hoooo,* while the contact call more commonly given by females is a sharper *kee-whik*. So if you hear a pair together, the male calls *to-whoo*, while the female dismissively replies *twit*!

WHEN AND WHERE TO SEE

All year round in woodlands and suitable parks and gardens.

Barn Owl

BARN OWLS ARE rural birds, named for their habit of nesting in and around farmland, often in barns. They favour lowland farmland and it's here that is best to watch for them hunting at dawn and dusk. They hunt by stealth, flying silently and methodically, quartering fields looking for voles and other potential meals.

Despite an owl's reputation for seeing in the dark, their eyesight is actually not much better than ours. Instead it's their hearing that enables them to pinpoint their prey (see page 35). Studies have shown that a Barn Owl can capture moving prey in almost total darkness, just through its prodigious powers of sound-tracking. This means that Barn Owls hunt best on relatively still, quiet nights, though a bit of a breeze is helpful to provide uplift and reduce their flight effort.

TOP ID TIPS

Thanks to their eerily silent flight and pale colouring, Barn Owls are sometimes nicknamed ghost birds. They have white heart-shaped faces and white bellies, with light golden-brown upper wings. In flight they are easily distinguished as an owl, with their large heads, broad, long wings and short tails. Most females are slightly darker on the underside than males, with some fine speckling.

The best chance of seeing a Barn Owl is at dawn and dusk, particularly in winter, but also when the birds have hungry young to feed, which is usually between May and July. Their call is an eerie screech.

WHEN AND WHERE TO SEE

All year round in lowland grassland or on farmland.

October's challenge bird

Short-eared Owl

SHORT-EARED OWLS are scarce birds in the UK, which breed mainly in northern England and Scotland. In the winter months, numbers are boosted by migrant birds from Scandinavia, Russia and Iceland, making it possible to see these medium-sized owls further south and also in Wales. Coastal marshes and wetlands are good places to seek them out. Unlike most other resident owls, 'Shorties', as they are affectionately known, can be seen hunting in the day. They are named for the small 'ear-tufts' on the sides of their faces. However, these tufts are tiny and rarely seen, plus they aren't even ears at all but short clumps of feathers. In Orkney, Short-eared Owls go by the name of 'catty face'.

TOP ID TIPS

Like Barn Owls, Short-eared Owls are most likely to be seen in flight, as they quarter grasslands and marshes looking for voles and other prey. But they are browner overall, with a smaller face than a Barn Owl.

They have long wings with dark tips and the face is round and unmistakably owlish. Look for their piercing yellow eyes, which can even be seen at a distance. Seen closer, the face is white, with fierce black 'make-up' accenting their bright yellow eyes. In winter you are unlikely to hear them call, but during the spring breeding season, males can be heard hooting softly.

WHEN AND WHERE TO SEE

Hunting over wetlands and coastal marshes in winter.

Monthly musings

Nutty for acorns

A Jay's appetite for acorns is enormous. Over the season, just one single bird can gather around 5,000 acorns, depositing them up to several kilometres from where it first found them. As well as carrying acorns in their mouths, Jays store them in their crop (a pouch in their throat to hold food) and gullet, enabling them to carry a gobsmacking nine acorns – although, presumably for comfort, three is the more typical load. Minimising the chances of losing their treasure, they hide them in numerous places. But with so many acorns and so many hiding places, a Jay will never retrieve them all. Instead some acorns will grow into oaks, providing more food for the Jays of the future.

Birding while you sleep

Migration is a risky business for birds. Not only do they need the sheer physical strength to make the journey, but they also have to avoid the hungry clutches of predators. One way to minimise the chances of being eaten is to migrate under the cover of darkness, and it's why many birds migrate at night. This has led people to 'nocmigging', the word being a mash-up of nocturnal and migration. It's an engaging hobby that essentially involves setting an audio device to record overnight. Any bird calls picked up are then identified the next day to discover which birds were on the move.

← Jays collect acorns in autumn.

↑ Pink-footed Geese flying to their roost.

Top two must-dos

1 Hear a Tawny Owl hooting

Nocturnal animals are inevitably difficult to experience and, being only active at night, Tawny Owls are no exception. However, from autumn you can get closer to these enigmatic birds by heading out to hear them hoot.

Noisy nights

Tawny Owls are territorial birds, and from autumn through winter they are keen to make their presence known. When a pair are interacting, the male's haunting *hu-hu-hoooooo* followed by the female's *kee-whik* is a duet designed to let the other Tawnies know exactly who owns what. There are new kids on the block too,

with last year's chicks, hoping to be next year's young couples, all keen to establish territories of their own. Although the birds won't settle down to lay eggs until around March, Tawny Owls have good reason to mark their patch early. These are highly sedentary birds, spending their whole lives in one stomping ground typically not far from their birthplace. Their intimate knowledge of just one place can help them to find food, particularly in times of scarcity. But this stay-at-home nature can also curtail their spread into other areas. Tawny Owls are not found in Ireland as, like other sedentary birds, they are presumably unwilling to fly over sea to colonise new countries.

↑ Tawny Owls are at their most vocal in autumn.

→ Redwings will forage on fallen apples.

But for those of us in Britain lucky enough to live close to woodlands, we can expect nights of noisy hooting, giving us a rare glimpse into the Tawny Owl's world. You can also hear them in urban areas, where they nest in trees and even old buildings, so long as there are sufficient foraging opportunities close by.

Flash mobs

Should you hear Tawny Owls nearby, there is always a chance that you might see one. In the daytime, listen for a mobbing by angry birds. Tawny Owls roost during the day, hiding away in the cover of a tree. But sometimes, somehow, their cover is blown, causing much consternation among the local smaller birds who will hound this potential killer, calling and flapping around it until the owl flies off. Of course it will be back when the fuss dies down, settling back in its territory and ready at nightfall to stalk the birds that tried to cast it out.

Come April through May, you may hear the raspy calls of chicks as they cry out for food. Keep an eye out for these balls of fluff, as they tentatively emerge at dusk. It's a short phase known as 'branching', when the young owls get their bearings, edging out into the trees, walking, climbing and sometimes jumping from branch to branch.

2 Welcome flocks of wintering thrushes

Every autumn the UK welcomes two wintering thrushes: hungry Fieldfares and Redwings drawn in by the berry bonanza unfolding across the country. Both species have very small UK breeding populations, but many thousands arrive from further afield in winter.

Listen out for migrating Redwings

Flying in from Iceland, Russia and Scandinavia are Redwings, the smallest thrush we can expect to see in the UK. These birds have an orangey-red tinge under their wings and on the flanks, which can be made out as the bird perches and flies. They also have a pretty white stripe above their eyes, adding a bandit-like edge to their looks.

Redwings arrive in the UK from October and can be seen throughout the

country, often in flocks with other birds, particularly Starlings and Fieldfares. Look for them in hedgerows or open ground in the countryside, and in parks and playing fields. As well as eating berries and fruit, Redwings eat earthworms and can be seen foraging on fields, particularly as berry supplies run lower as winter progresses. Listen out for their high-pitched *seep, seep,* as they call to each other frequently.

These are also birds that you can experience on migration. They fly here by night, calling constantly with the same *seep, seep*, giving us a chance to hear them as they migrate in their thousands. Wherever you live, you could hear them overhead. Simply step outside on a still night in October, or open your window and listen.

Feast your eyes on Fieldfares

Next up is the Fieldfare, the UK's largest thrush, flying in from Finland, Norway, Sweden, Russia and eastern Europe. Sometimes mistaken for a Mistle Thrush, the Fieldfare has a grey head and yellow bill, quite unlike the brown heads of other thrushes. The wings are red-brown, and it has a grey rump and a fairly long black tail. Just like Redwings, Fieldfares can be seen in flocks up and down the country, in playing fields or open country, where there are hedges and trees nearby.

Look for Fieldfares foraging on the ground for worms and other insects, as well as in hedges and on trees, picking at berries. Their call is a short *chack, chack* which can sometimes sound like a chuckle, and they too can be heard on night-time migration as well as during the day.

Fix a feast for hungry guests

As winter draws on and berries dwindle, your chances of tempting winter thrushes to your garden increase. Apples, especially, and other fruit can attract both Fieldfares and Redwings, particularly in times of prolonged cold or snow.

↑ Fieldfares and Redwings can often be seen together.

How to help

Celebrate World Migratory Bird Day

The seasonal mass movements of birds are remarkable, with our winged wonders facing numerous challenges as they navigate their way around the globe in search of better food sources and breeding grounds. It's a global phenomenon and a reminder that, as wildlife moves between countries, efforts to protect it are necessary at a worldwide scale. World Migratory Bird Day is an annual celebration of these efforts.

Bird migration peaks at two times of the year, coinciding with spring and autumn in the UK. This means there are two World Migratory Bird Days each year, and these are held on the second Saturday in October and the second Saturday in May.

Across the world people hold events to celebrate these days and the RSPB, the Wildlife Trusts, the Wildfowl & Wetlands Trust and others host activities that can help you get to know more about migratory birds and the ongoing conservation needed to protect them.

The first World Migratory Bird Day was celebrated in 2006, following earlier initiatives celebrated in the United States. The events are now overseen by the United Nations Environment Programme and are themed to highlight different aspects of the challenges faced by migratory birds. These have included plastic pollution, illegal killing, deforestation and climate change. Search 'Migratory Bird Day' online to find out how you can get involved.

↓ Wading birds gather in huge numbers at RSPB Snettisham in winter.

Plant berry-bearing shrubs and trees for birds

The best time to plant a fruit or berry tree or bush in the garden is from October to April, and a little bit of effort now could reap huge dividends for birds for many years to come.

As well as providing food for birds, other wildlife can also benefit. Insects such as bees and butterflies may feast on the nectar-rich flowers, while mammals including Hedgehogs, Badgers, mice and even Foxes will eat berries. The trees and bushes can also provide shelter for wildlife, and you may find birds choosing to make their nests in them.

Trees and shrubs that are native to Britain are often better than non-native species. This is because our wild fauna has evolved over time with the plants, making these plant species better at supporting our native animals.

Another great plant for wildlife year-round is Ivy. Unlike most other plants, Ivy flowers in late autumn and doesn't bear berries until midwinter. It's a great source

of food when others are drying up. The berries are eaten by a whole host of birds including Blackbirds, Starlings and Jays, as well as migrant Redwings, Fieldfares and Waxwings.

TOP TREES FOR GARDENS

The best fruit-bearing trees to help your garden wildlife include:

- Crab Apple (*Malus sylvestris*)
- Rowan or Mountain Ash (*Sorbus aucuparia*)
- Spindle (*Euonymus europaeus*)

Others, which can be planted as a specimen tree or as hedges, include:

- Yew (*Taxus baccata*)
- Blackthorn (*Prunus spinosa*)
- Juniper (*Juniperus communis*)
- Hawthorn (*Crataegus monogyna*)
- Holly (*Ilex aquifolium*)

↓ A young male Blackbird enjoys a Rowan berry.

Myth of the month

Birds of wisdom and woe

Owls have long been considered wise in Western cultures. In ancient Greek mythology, a Little Owl represents the goddess of wisdom and war, Athena, and this is reflected in the modern-day scientific name for the bird, *Athene noctua*. To see an owl was considered good luck as this was a sign from Athena, and even today we refer to 'wise old owls'. But do owls warrant their wise reputation?

One measure of intelligence is problem-solving, but it's members of the crow and parrot families that come top of the class here and not owls. Owls are, though, very successful at what they do, with their sharpened senses and natural adaptations that help them hunt at night. So while they might not be classically wise, they have what they need to survive.

The nocturnal nature and often eerie calls of owl species such as Tawny and Barn Owls have also led to a more sinister association in British folklore, with the birds often considered to bring death and misfortune. In Shakespeare's *Richard III*, exasperated by reports of bad news, the king chides the messenger, saying 'Out on ye, owls! nothing but songs of death?' In the mythical Welsh stories of the Mabinogi, the wizard Gwydion creates a beautiful woman, Blodeuwedd, out of wildflowers, to be wife of Lleu Llaw Gyffes. But she takes a lover and as punishment is transformed into an owl, never again to show her face in daylight. With its haunting cries and ghostly beauty, it's not too hard to think that a Barn Owl may have inspired this tale.

↓ Little Owls are associated with wisdom.

11

NOVEMBER

Flock together

This month is a time when birds gather together, seeking out the best foraging grounds, huddling together for warmth and teaming up for safety. For us birders, it's a fantastic opportunity to see huge flocks and watch dramatic displays.

SPY A BIRD in November and chances are you'll see a flock. This is the season to come together ahead of colder weather and team up to find food. Just as many people will be gathering to watch the glittering flashes of fireworks in the night sky, there's an equally mesmerising dusk-time display as thousands of Starlings flock before bedtime. November is a time to brave the damp, and head out to enjoy nature's displays.

With the busy task of parenting over, all rivalries are set aside as many birds flock together in autumn. Starlings form famously big gatherings, with UK residents joined by birds from the continent for the winter. Lapwings noticeably build in numbers, too, as birds from colder climes join them. Some birds also team up with other species. Tits form groups with other tits and are often joined by other small birds. Spot a Great Tit and, among the throng, you may also see Blue Tits, Long-tailed Tits and even tiny Goldcrests as well as Treecreepers and Nuthatches.

You can also see great groups of finches, with Chaffinches joined by Greenfinches

← Fieldfares are migratory birds that winter in the UK.

↑ Goldfinches are frequently seen in flocks.

and sometimes by Bramblings. Closely related to Chaffinches, these migrant finches fly in from the frozen north to winter in our relative warmth. You may also see Blackbirds, Mistle Thrushes and Song Thrushes foraging with their visiting cousins from Scandinavia and beyond, Redwings and Fieldfares.

Crowd appeal

From our parks and gardens to our open countryside and wild woodlands, it seems all the birds are gathering, everywhere. There are, however, some notable exceptions to this rule, with Robins and Wrens staying fiercely independent. But for others, the crowd appeals, and it's likely to their advantage. As us humans say, 'There's safety in numbers', and it's a maxim these birds stick to in winter. Birds risk predation from various other animals, including birds of prey such as Sparrowhawks and mammals such as Foxes. In autumn and winter, when they have to spend a lot of their time feeding, they are frequently vulnerable. But if a bird feeds in a flock, there are more eyes to spot predators. There are also more birds for a predator to eat, so foraging with others could lessen an individual's chances of being eaten. It's a similar story when birds fly in flocks. With so many moving targets, it can be harder for a raptor to single out and pick off one individual.

Birds may also get together to share information about good places to feed. More eyes mean more chances of finding good spots. Similarly, it may be that where one bird has found a rich source of food, the rest of the flock can follow. Moving together has another practical advantage for finding food. Mixed flocks of tits and other small birds move through woodlands

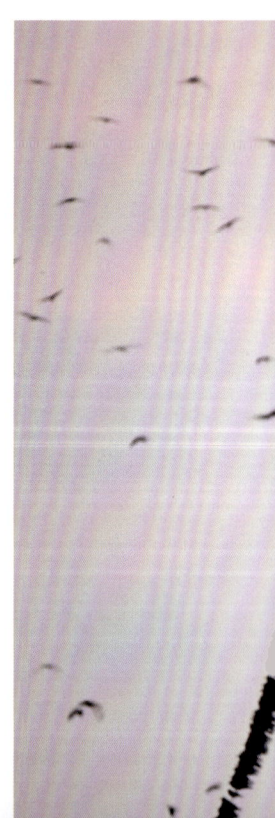

← Long-tailed Tits and Blue Tits take advantage of a feeder.

→ In winter, Rooks gather before roosting together.

or hedgerows feeding on insects, spiders and other invertebrates. As the group descends, the many eyes and bills will ensure few can escape.

Long, cold nights

It's not just during the day that birds come together. As well as feeding up to survive the night, birds need somewhere safe and warm to sleep. Typically smaller birds retreat into the shelter of hedgerows and trees, with tree holes and nest boxes also favoured by birds such as Great Tits, Wrens and Blue Tits. But should the weather get cold enough, some may pass on their personal space and huddle with others to survive. This can be extreme in the case of the Wren (see page 164), with a record of 61 individuals seen entering one nest box on a very cold winter's night in Norfolk in 1969.

Carrion Crows and Starlings are two more birds that roost in large numbers. Starlings roost together in their hundreds and thousands, performing captivating murmurations where they swirl together in great black clouds, before settling down for the night. Carrion Crows, meanwhile, often get together with other members of the crow family to roost, joining with Jackdaws and Rooks to sleep together in tall trees. These roosts can be huge, holding thousands of birds. As the birds gather, the sky can darken with the flapping of black feathers and the air fill with the noise of chattering corvids. Ravens, the largest of the crow family, also roost together, with a roost site in Anglesey once attracting 2,000 birds. Raven roosts are lively affairs, with the birds calling and soaring in pre-roost displays, before re-settling.

← Bramblings migrate to the UK when conditions are harsh in their homelands.

Better together?

Quite why birds form such large roosts for sleeping is not fully understood. As well as gathering for warmth, the advantages of communal roosting could be similar to those of daytime flocking. Being part of a larger group may mean there is less chance of being predated. Certainly predators will notice such a large roost of birds, but with such a mighty banquet on offer the individual's chances of being eaten are lower.

However, for large birds such as Ravens and Carrion Crows, a fear of predation is less likely to be the motivation. Building social bonds and perhaps finding a mate may be more important. One theory is that roosts provide birds with an opportunity to learn from each other, particularly when it comes to finding food. During the day birds break up into smaller flocks. When they return to the roost, others pick up on clues to how well their foraging went, and those deemed successful are followed the next day to their feeding grounds.

A study of the Anglesey Raven roost by researchers at nearby Bangor University supports a theory of information sharing. The researchers used sheep carcasses baited with colour-coded digestible beads. By analysing the beads found under the roost, they discovered that within the roost the Ravens formed small groups. The birds within these groups roosted and ate together, sharing information about food sources with each other.

Enjoy the show

In the shorter days of November and further into winter, we may fear that the lack of daylight hours will decrease our opportunities for birding. But the early nightfall can make it easier to see birds heading to roost, and in the case of birds such as Starlings and Carrion Crows this provides a great chance to witness their intriguing pre-roost displays.

Birds of the month

Starling

STARLINGS ARE INQUISITIVE little characters who descend all at once, squabbling, strutting and stuffing themselves. They are noisy birds, frequently chatting to each other with buzzes and wheezes. They are also incredible mimics and show off these skills as part of their repertoire when singing. From high on a tree or a chimney, they can begin with gliding wolf whistles, turning to a rapid popping noise that burbles with increasing intensity. Then they turn copycat, calling softly like a Red Kite, mimicking a car alarm, chirping like sparrows and throwing in a barking dog for good measure. Sadly, these talented birds are now on the Red List of Birds of Conservation Concern, with the numbers breeding in the UK down 53 per cent between 1995 and 2020.

TOP ID TIPS

Starlings are slim, upright birds with shimmering black feathers. In spring and early summer, when they are out to impress, the green and purple iridescence is at its most prominent. In winter, their feathers are sprinkled with tiny white dots, the 'stars' that inspired their name, and perhaps a festive nod to the season of snow. Males and females are hard to distinguish, but in spring and summer you could be in with a chance. Look closely at their yellow bill; if the base is bluish-grey it's a male. If it's a palish pink, then it's a female. Juveniles are plain pale brown, with black bills.

WHEN AND WHERE TO SEE

All year round in gardens, parks, woodlands and the wider countryside.

Carrion Crow

THESE ARE FAMILIAR birds, distinctive with their all-black feathers, dagger-like bills and harsh cawing call. Carrion Crows can be quite shy when it comes to humans, perhaps a legacy of hundreds of years of persecution. They are infrequent visitors to gardens but are often seen in parks or the countryside where they walk with a swaggering gait. They are omnivores and opportunists, as at home eating roadkill as they are foraging for berries and grain. Seafood, too, is on the menu for those by the sea. Here they will fly high with crabs or molluscs, dropping them onto rocks to break them so they can eat the soft insides.

↓ Hooded Crows (inset) replace Carrion Crows in north-west Scotland and Ireland.

TOP ID TIPS

Carrion Crows can be confused with other members of the crow family. However, their bills should set them apart from Rooks. A Carrion Crow's bill is black all over, while a Rook's looks whitish-grey, and longer (thanks to the bare patch around its base). Jackdaws, too, can cause a challenge, but these are smaller birds, with much shorter bills. To help tell them apart, listen for the Jackdaw's lively *jack, jack, jack* call, which is very different from a Carrion Crow's harsh *krrah*. Close up, Jackdaws have grey neck feathers and pale blue eyes. In north-west Scotland and Ireland, Carrion Crows are replaced by Hooded Crows. They are closely related but easily told apart as the 'Hoodie' has a grey body with black head and wings.

WHEN AND WHERE TO SEE
All year round in all habitats.

Brambling

THESE PRETTY FINCHES are winter visitors that come to the UK when the weather turns cold in Scandinavia and Russia. They come here to escape harsh weather and a lack of food, and the numbers turning up in the UK vary according to how severe the weather is in their homelands.

Bramblings are close cousins of Chaffinches. The two are very similar in size and shape and are often found in flocks together. These birds have a taste for beechmast, the triangular nuts in spiky brown cases produced by Common Beech trees. Bramblings can be seen anywhere in the UK from autumn onwards, though numbers vary a lot year on year. Most birds will have left by April.

TOP ID TIPS

The easiest way to tell a Brambling from a Chaffinch is to look for a brilliant white rump in flight. Both male and female Bramblings share this characteristic, and this marks them out as not Chaffinches. In winter, male Bramblings have an orange breast and whiter belly, with a brown-black head and wings. The female is similar but with paler orange tones, and the head is greyer too.

It may also be possible to tell Bramblings from Chaffinches from their call as both birds call frequently. When feeding on the ground, a Chaffinch calls with a repeated *pink, pink, pink*, while a Brambling has a buzzy *te-eep*.

WHEN AND WHERE TO SEE

Countryside and woodlands, from October to April.

Fieldfare

FIELDFARES are large thrushes that arrive on the east coasts of the UK from October before moving steadily inland as the season progresses. These birds from eastern Europe, Scandinavia and Finland fly in for our relatively milder winters. Fieldfares eat berries and the numbers migrating here each year depend on the berry crops in their homelands. Migration uses immense amounts of energy and there are risks along the way, so if there is food available at home, the birds will stay rather than fly thousands of miles elsewhere.

Since 2009, the Fieldfare has been included on the Red List of Birds of Conservation Concern in the UK, as a result of a very small and declining breeding population. While wintering flocks can be found almost anywhere in the UK, the largest numbers are typically seen in central and eastern England.

TOP ID TIPS

Fieldfares can be seen with other thrushes, picking out berries in hedgerows or in flocks on trees and fields. Like other thrushes, they have speckles on their breasts and flanks. However, they are more colourful-looking birds, with blue-grey heads and rumps, and black tails. They have brown wings, and in flight the underside of the wing is bright white. Male and female are alike. Young birds, experiencing their first winter, can be made out by their less orangey breasts. Fieldfares make a short chucking *chack, chack* call.

WHEN AND WHERE TO SEE

Open countryside, parks and woodland edges, from October to April.

Redwing

REDWINGS ARE THRUSHES that migrate to the UK from late September. Like Fieldfares, they spend the winter here to avoid tougher conditions further north or east, and to feast on the berries abundant in the countryside. As with Fieldfares, the numbers arriving vary according to how bountiful the berry crops are at home, with a poor harvest meaning more arrivals. Redwings were included on the Red List of Birds of Conservation Concern until 2021, when their status was changed to Amber, thanks to an increase in their small breeding population in the UK.

TOP ID TIPS

Redwings are the same size as Song Thrushes and easily confused, particularly as the red under the wing can be difficult to make out in poor winter light and from a distance. They also have brown backs and wings, with a speckled belly that can look like a Song Thrush's. However, the Redwing's spots tend to be joined up into streaks. If you see a flock of thrushes foraging on the ground in winter, one way to tell them apart is to look out for the Redwing's bold white stripe above the eye, as it can often seem prominent against the bird's otherwise brown head. Fieldfares also have a white stripe, but it is shorter and fainter and their head is otherwise grey.

WHEN AND WHERE TO SEE

Open countryside, parks and woodland edges, from September to April.

November's challenge bird

Raven

BIRDS OF MYTH and magic, Ravens occupy a special place in British folklore; should the birds leave the Tower of London, then the crown and kingdom will fall, and so a Ravenmaster is employed to protect several captive Ravens there (see page 129). Excepting these tame birds of the Tower, Ravens are generally shy and not commonly seen. They have also suffered through persecution and habitat loss, but more recently their fortunes have changed. Ravens are now on the up, strong in their mountainous Welsh homelands and expanding in range in England and Scotland.

Sighting a Raven can be magical, as a giant black shape casts its shadow on the ground below. Often in pairs, they fly with strong, measured wingbeats, soaring over huge territories, calling gruffly to each other. These are loyal birds that pair for life, with an intelligence likened to that of gorillas.

TOP ID TIPS

Although Ravens are huge birds, as big as a Buzzard, they can still be confused with Carrion Crows, especially in flight when it can be hard to gauge size. The clearest sign is to listen for their loud *kronk*, which is deeper than a Carrion Crow's *krrah*. Look out too for a diamond-shaped tail, long wings and finger-like wingtips. They have heavy bills, larger than a Carrion Crow's, and a thick mass of feathers around their necks. They're also fond of aerobatics, and you may see them tumbling and rolling overhead.

WHEN AND WHERE TO SEE

Strongholds in upland areas and coastal cliffs across the UK but increasingly common in lowland farmland and forestry too, all year round.

Monthly musings

Say it with gold

Goldfinches are charming birds with red faces and bright yellow wing-bars. But it's thought that these colourings are for far more than beauty. One theory is that the yellow markings help the birds to communicate. When a Goldfinch takes off, its wings flash, warning other birds of danger. The wing markings can also help the birds stick together when flying, with the highly visible yellow making them easily identifiable. Another theory is that the colouring is a warning. Much in the way that yellow and black warn predators of stinging bees and wasps, the yellow and black might act as a deterrent, warning that the bird is not good to eat, although Sparrowhawks and cats have clearly not got the memo.

Ruling the roost

You may think that a night-time roost is a cosy affair with all birds keeping close for warmth. However, there is a pecking order. Researchers have found a hierarchy within Rook roosts, with the older dominant birds taking the top spots. Usually these are found towards the top of the tree, where the birds can sleep without fear of faeces falling on them through the night. However, if the weather becomes harsh, the dominant Rooks will move further down the tree, forcing the younger Rooks out and into less sheltered parts of the roost. We may baulk at the thought of adults forsaking the young to protect themselves, but perhaps it's a strategy to ensure the experienced adults can breed again, come spring?

↓ Rooks can gather in great numbers.

↑ Goldfinch in flight.

Top two must-dos

1 Be mesmerised by a Starling murmuration

A Starling murmuration is one of nature's most amazing spectacles, when just before sunset thousands of birds swirl and dance in the sky, dense as dark clouds, twisting and surging into myriad shapes. The term 'murmuration' is thought to be derived from the murmuring sound heard as the birds flap their wings overhead, a reminder that this is a show that's as much a feast for the ears as for the eyes.

Starlings murmurate in the winter months, as they gather before settling in their roost for the night. One suggestion for the murmuration's purpose is that it advertises the roost, telling others where they are gathering. A roost is a warmer option than sleeping alone as the body heat from so many thousands of birds warms up everyone. Another reason for their mesmerising moves is for safety. The murmuration makes it harder for a hungry raptor to pinpoint a kill. Flying into the roost as one of thousands, rather than one alone, lessens the odds of you being somebody's dinner.

Murmurations typically involve thousands of birds, and they grow in numbers as winter progresses and more birds arrive from colder parts of Europe. The colder it gets in eastern Europe, the bigger the size of our flocks. The biggest European murmuration is found in Rome, where a 10 million-strong 'super swarm' flocks above the city at night. Despite

↓ Watching a Starling murmuration in Brighton.

→ Starling murmuration at RSPB Minsmere.

flying in such huge numbers, Starlings don't crash, thanks to reactions 13 times faster than ours. With remarkable spatial awareness they focus on their nearest six or seven neighbours, flying in synchrony with them.

Top tips for watching a murmuration

The peak time to see this stunning display is the beginning of November until the end of January, but they start as early as October and can continue into March. Roost sites change from year to year, but good places to watch include Aberystwyth and Brighton piers, and numerous RSPB nature reserves including Burton Mere Wetlands in Cheshire, Mersehead in Dumfries, Newport Wetlands in south Wales and Portmore Lough in County Antrim. Discover more sites by searching 'Starling murmuration' online.

To watch a murmuration, head to the site around an hour or so before sunset, so you've more chance of seeing the show. It typically takes place before sunset and lasts around 20 to 30 minutes, but timings can vary and sometimes it may not happen at all. Nature is nothing if not unpredictable!

2 Seek out a city roost

From November onwards, look out for Pied Wagtails roosting in city trees as night falls. As temperatures drop, birds cluster together, taking over a tree and looking like baubles at Christmas time. It's a particular thrill as, with little regard to the busy town around them, these birds gather noisily at dusk, alerting others to the roost with their sharp *chizzik* calls.

Pied Wagtails are common birds seen year-round, bobbing their long tails as they walk along. They are dapper grey, black and white birds with white bellies and long black tails. They can be seen anywhere but are often encountered in towns and cities, strutting along the pavement, flitting in and out of the road, seeking out insects to eat. The jury's still out on the purpose of their incessant tail-bobbing. One suggestion is that it helps flush out their prey. Another is that it helps camouflage their movement when they are hunting by running water. But it could also act as a signal, alerting other

wagtails to its social status or warning predators that this bird is alert and not easy to catch.

Where to see a city roost

As the weather grows colder, Pied Wagtails roost together at night, and as our urban areas are a few degrees warmer than the countryside, a tree in a town or city makes a good choice. Sheltered sites such as trees planted near hospitals, supermarkets or bustling city centres can prove fruitful places to look. They'll also roost on artificial structures such as service station roofs. Surprisingly, Pied Wagtails often pick busier, brighter places and have been known to roost in Christmas trees. Perhaps these are warmer or feel safer from predators.

You're most likely to find a city roost in southerly areas as, although Pied Wagtails breed throughout the UK, in winter many move south, avoiding the harsh winters of upland areas of Scotland and northern England. Numbers are also swelled by migrant birds joining our resident birds from colder parts of Europe.

Under the night-time glow of streetlamps, Pied Wagtails can look like white lollipops, so it's not surprising that they can be misidentified as Long-tailed Tits. However, Long-tailed Tits favour more private roosts and don't gather in such great groups.

Not just city birds

Pied Wagtails also roost in the countryside, with a preference for reedbeds. Reedbeds are also a favourite of Starlings. The water below the reeds provides a great natural defence against predators such as Foxes. Plus, the water can help keep the temperature of the reeds a little bit higher than that of the surrounding countryside.

↑ Pied Wagtails often roost in towns and cities in winter.

How to help

Be a little bit wild

Autumn can seem like the ideal time to get busy in the garden, tidying up overgrown branches, removing rogue growth and clipping back unruly bushes. However, for birds and other wildlife looking for food and shelter, what we see as mess may be just what they need.

One of the easiest things to do is to leave seedheads on plants.

Plants such as Teasel, echinacea, Globe Thistle, rudbeckia and alliums are good choices. Teasels and thistles are favourites of Goldfinches. Keep an eye out for small flocks flying in, nimbly perching on the plants as they tug away at the seeds. It makes for a great photo opportunity. As well as providing seeds for birds, the stems often become a home for wintering insects.

Ivy is a particularly valuable plant, as by bearing berries from November to January it provides food for birds right through the winter. Its gloriously thick tangle of evergreen foliage also means that it's a firm favourite among garden birds looking for a place to sleep. Wrens, Robins and House Sparrows are happy to hide away here. Insects, too, often hibernate in Ivy.

When it comes to trimming hedges or trimming back trees, you may also want to think about leaving some shelter for the wildlife. But what you do with the cuttings can be just as valuable. One way to help insects and other invertebrates is by making a log pile house, which is

← Woodpigeon eating Ivy berries.

↑ Goldcrest foraging.

a rather grand way to describe leaving a clump of woody cuttings in the garden to decompose naturally. This is invaluable to insects, and of course, by looking after the insects you're also looking after your birds, as many are insectivores.

Join a community of bird-lovers

Amazing things happen when people come together. Up and down the UK, there are many community groups working for nature and making a huge difference for wildlife. The RSPB has a network of more than 130 local groups, all doing a whole host of things, from organising bird walks and talks, to carrying out wildlife surveys or putting up nest boxes. Some have even created new habitat: in North Ayrshire the local group helped to create an island in an artificial lagoon to make a nesting place for gulls and Sandwich Terns.

By being part of a local group you're not only helping wildlife, but you've also found a great way to meet like-minded people and discover more about your local birdlife. Many run social events and outings to local nature hot spots as well as nature reserves further afield. You can find details of local groups near you on the RSPB's website.

The Wildlife Trusts also have a network of local groups for people to get together to enjoy and help wildlife. As well as organising nature walks and talks, groups also help out on nature reserves and fundraise for nature.

And don't forget the many independent community groups that look after nature. Try looking on social media or online. There's sure to be a group waiting for you.

↓ RSPB Medway local group carrying out a litter pick.

↑ Ravens are playful birds and display by flying upside down (below).

Myth of the month

Blowing in the wind

There is a lot of weather lore around birds, with one such being that if Carrion Crows or Ravens are flying at a great height then the winds are strong. Of course in stormy or blustery weather, we're probably well aware it's blowing a hoolie, whether or not we can see corvids high in the sky above. This is one of those sayings that confirms what we already know rather than tells us something we don't. But it does indicate a happy truth. Ravens and Carrion Crows do play in the wind. Head to the coast or up a hill on a windy day and watch for these feathered surfers. As if catching a wave, you can see them flying into the wind then gliding as it carries them away before they flap back down.

Ravens have a particularly cool party trick and will fly upside down, wings half-closed, as they tumble down before turning. From spring, pairs will perform these acrobatics as courtship, but it appears they also do it just for fun. Snow, too, can bring out the playful side of these fun-loving birds, with Ravens and Carrion and Hooded Crows seen playing in the snow, rolling or sliding down slopes.

Play in young birds and other animals is thought to help them gain the skills and coordination needed for adult life, such as for hunting. Daredevil aerobatics are core to a Raven's courtship and pair bonding so it's not surprising these birds are keen to test their new moves. But however much we rationalise the reasoning behind our wild birds' behaviour, there's no getting away from the sheer joy sparked by watching a Raven falling and spinning in the sky.

12

DECEMBER

Christmas quackers

December feels like a quiet time of the year, and in terms of bird movements, it is. But the UK is still bustling, with enormous numbers of birds everywhere, from fields to lakes and from estuaries to marshes. It is a good time to get out and see them.

THE LAST MONTH of the year sees birds' migratory movements decline as most are now settled into their winter quarters. That still leaves plenty to be seen, though. There are thousands of wildfowl and waders on our estuaries and marshes, and it's a great month to enjoy grand-scale birding, with everything from Starling roosts to skeins of geese. You need to fit it all into the short days, though.

The next few months see the highest numbers of wildfowl in the UK. Even the very last regular arrivals, Bewick's Swans from Russia, have arrived and settled in. Just about every pond and lake will hold birds that have travelled from elsewhere to winter here, even if it's just Mallards on your local park pond that have come from Sweden or the Low Countries. If December brings harsh conditions, with long-lasting heavy frosts and snow, birds will be displaced within the country and many more will cross over from the continent, adding to the crowds.

It's a wonderful time to enjoy ducks. The males are in their breeding finery, with bold patterns, fancy ornamentations and colourful accents. The females are the epitome of subtle speckling and barring. There is much interplay between the two. What looks like birds shaking their heads or flapping their wings isn't happening to get comfortable; it's subtle display. Watch any group of male Mallards

→ Look out for the 'head-throwing' display of the male Goldeneye.

← Many Mallards that you see in December are winter visitors from the continent.

and you will notice them abruptly lifting their rear ends and dipping their heads down to touch the water, probably making a weak whistle as they do so. This is courtship, and at this time of the year it will be communal, with several males displaying to apparently disinterested females. You will often see what are called 'three-bird flights', in which a female suddenly takes off, pursued by two (or more) males, all quacking. This is also display. It doesn't look that different from everyday behaviour, so you can easily miss it.

The ducks that winter on the sea, which are indeed often called 'seaducks' by birdwatchers, often have more ostentatious forms of courting. Male Red-breasted Mergansers have a wonderful bowing and 'sky-pointing' display, while Goldeneyes throw their head back so that it rests on their back. Eiders also throw their heads back, while uttering the most gorgeous, deliciously suggestive cooing, which will make you smile every time you hear it.

Divers and dabblers

Watching the ducks on a pond, lake or estuary gives you an excellent chance to appreciate the differences between their lifestyles. On the sea, things are simple, because every duck is a diving duck. Eiders and Common Scoters dive underwater to snatch shellfish, such as mussels, from the seabed, whereas mergansers dive underwater to chase fish.

In freshwater, the two commonest diving ducks are Tufted Duck and Pochard, the former feeding on a substantial amount of animal matter, and the latter veering more towards a vegetarian diet. Pochards have a habit of feeding at night, so you often see them in rafts asleep during the day. Most other ducks

← Upending is a common feeding method of the long-necked Pintail.

↑ In display, the male Eider (front) makes a delicious, saucy coo.

are described as 'surface-feeding' or 'dabbling' ducks, which prefer not to dive underwater, but appreciating their slightly different feeding habits is enjoyable. Notice, for example, that Teal mostly feed at the very edge of water, just about getting their feet wet. Shovelers seem to spend most time keeping their bills at the very surface of the water while swimming – classic dabbling. Gadwalls are experts in the art of simply dipping their head into the water while swimming, while Wigeons, if they can, forsake the water altogether and feed on land, nibbling grass in the company of geese. Pintails, with much the longest necks of any ducks, do what you might expect – they upend, dipping their whole forebody into the water, with rear end pointing skywards, so that they can reach deeper under than the rest.

All these species are able to use everyone else's methods, and it also depends on water level and conditions, but most have a preference. The exception is the remarkable Mallard, which can do everything.

Cold, damp and dark

Ducks seem immune to the cold, damp days of December. They are powerful flyers that can evacuate an area easily, and they can feed day or night, often doing the latter by the light of the moon. Many other birds are not so flexible. Robins and Wrens hold winter territories that they cannot forsake. Nuthatches, Jays and Willow Tits have territories full of winter stores that they have spent much of autumn gathering, and a single Mistle Thrush or Fieldfare may defend a berry-bearing tree (see page 33). Many birds take advantage of fixed feeding sources, from hedgerows stacked with

↓ Woodpigeons ruffle their feathers when it's cold.

↗ The Ptarmigan has feathered feet to reduce heat loss.

winter fruits to bird feeders, and cannot risk trying their luck elsewhere.

Survival on these short days depends on finding enough food in daytime, so that the fires within are fuelled for the long night. Feathers have fabulous insulating properties, and so long as the plumage (freshly grown after the autumn moult) is intact and well maintained, it should suffice to keep birds warm enough to survive. But this depends on the food supply not failing.

On cold winter days you will often see birds looking fatter than usual. This is because they are 'fluffing' their feathers to help them keep warm (see January, page 8). Robins often demonstrate this, as do pigeons.

Even in midwinter, most birds roost alone, but the very smallest-bodied species sometimes ruffle up together in bodily contact, ensuring that the heat is shared. Long-tailed Tits, Goldcrests, Treecreepers and Wrens all do this.

Snow White

Some birds seem to be immune to midwinter weather. Ravens, Hooded Crows and Golden Eagles use adaptability, guile and toughness to see any conditions out. In the lowlands, House Sparrows chattering in the forsythia seem to spend more time talking than eating. Capercaillies sit out the winter months up in Scots Pine trees, eating the needles.

The Ptarmigan, residing on Scottish mountaintops, is the toughest of all; it winters deep into the Arctic Circle. Its feet are fitted with feathers for insulation and to act as snowshoes, and it has thick layers of down beneath its outer plumage. It ekes out a meagre living of twigs and leaves. In the autumn it moults until the plumage is entirely white, for safety from Golden Eagles. Amazingly, it selects snowfields as the ideal place to roost at night.

Birds of the month

Mallard

THIS IS BY FAR our commonest, most widespread and most familiar duck, found everywhere from estuaries to village ponds. It is enormously adaptable, eating a wide variety of foodstuffs from grain to newts. Where people are still allowed to feed ducks, it is often the tamest and greediest species. It has long been domesticated, and this means that, over the years, various breeds have arisen that often escape and mix with the wild birds. So, there are many confusing variations, with different colours and shapes – some, for instance, are entirely white.

Mallards practise every main feeding technique, including upending and dabbling. They feed at the surface but will dive if trying to escape a predator. They are fast and powerful flyers and make a whistle with their wings as they go.

TOP ID TIPS

The male is handsome, although not always appreciated as such. He has a yellow bill, bottle-green head and purple-brown breast, while his tail bears very distinctive, curly black plumes, which are obvious in flight.

The rest of the body is greyish. The female and juvenile are mottled brown, the hue varying somewhat. All show wings with a white-bordered purple bar (the speculum). Both sexes quack, the female very loudly and in a series.

WHEN AND WHERE TO SEE

Abundant everywhere all year round. Large numbers migrate to the UK from abroad in winter.

↓ Both male (top) and female Mallards share the purple, white-bordered wing-bar or 'speculum'.

Tufted Duck

THE TUFTED DUCK is a professional diving duck that is largely confined to freshwater. A common inhabitant of park lakes, gravel pits and reservoirs, it is the most familiar submerging duck. It feeds from the bottom, obtaining a wide variety of animal foods such as freshwater mussels (a particular favourite), crustaceans and insect larvae, plus some plant material, especially seeds. Alongside the Mallard, it is the tamest duck, regularly coming to the provision of bread or grain, which it will sometimes submerge to find, so you can watch it swimming underwater, propelled by its back-set feet. Having feet so close to the rear end makes taking off difficult, so Tufted Ducks tend to have to run across the water before getting airborne. In common with all our ducks it is highly sociable, and often forms mixed flocks with its relative, the Pochard.

TOP ID TIPS

Small, with a rounded head, atop which sits anything from a substantial droopy crest to a bump. The eye is vivid yellow and the bill mainly blue-grey. Males are boldly patterned, black with gleaming white side-panels and a long crest, which hangs down behind the nape. Females and juveniles are brown versions of the same but are quite variable. Males in display make a naughty giggle, while females growl.

WHEN AND WHERE TO SEE

Common on large ponds, lakes and other freshwater habitats, sometimes large rivers. Widespread, but absent from higher ground. It is much more numerous in the winter.

↓ A male (behind) and female Tufted Duck.

↑ A male (behind) and female Shoveler.

Shoveler

IT'S HARD TO LOOK beyond a Shoveler's huge bill, both in identification and in guessing its feeding method. Among our ducks, it is the definitive filter feeder, sieving small edible items, both animal and vegetable, out of the water. Its bill is fitted with comb-like structures on both cutting edges. These are known as lamellae and many ducks have them, although they are particularly long and dense in Shovelers. These lamellae overlap to form a mesh. The Shoveler's bill is open-ended, so as it swims forward, water flows in. The water is then squeezed out sideways by the tongue, through the mesh, trapping particles (some of which, hopefully, are edible). Shovelers thrive especially in shallow, muddy water, and often purposely swim behind their peers, filtering mud stirred up by the feet of the bird in front.

TOP ID TIPS

The outsize bill always makes this relatively small and stocky duck easy to identify, and that's especially helpful for the females and juveniles, which have very similar patterning to other female ducks, although with more of a reddish-brown hue. The handsome male has a bottle-green head, prominent white breast and chestnut sides. In flight, both sexes have beautiful light blue forewings. Often takes off with a loud rattling of wings, but it doesn't call much.

WHEN AND WHERE TO SEE

A scarce breeding bird in shallow marshes or meadows, but quite common in many freshwater habitats in winter, from marshes and reservoirs to modestly sized undisturbed ponds. Often in flocks.

Teal

OUR SMALLEST DUCK, the Teal has a habit of 'springing' up from the water when disturbed, and it often occurs in quiet corners of lakes and marshes. It flies with fast wingbeats and its compact flocks often resemble those of a wader. It is sociable in the winter and, even when you cannot see Teal, you can often hear the males making their high-pitched, morse-code blips, while the females make the sort of quack you might imagine hearing from a Mallard holding its nose. Teals are inveterate dabblers, spending almost 90 per cent of their feeding time in water no deeper than 12cm. They typically paddle but can easily upend as well. They have a broad diet, but seeds are usually a main component.

TOP ID TIPS

It always looks diminutive in company with other ducks. It has a distinctly short neck and a small bill. The male is very distinctive, with his reddish-brown head split by a green streak, surrounded by a line the colour of gold plate. The yellow under the tail is easy to see at long range. Females and juveniles are mottled brown. Both sexes have an iridescent green panel in the wing (the speculum). Males make a sonar-like blip.

WHEN AND WHERE TO SEE

Common and widespread, with a small breeding population (5,000 pairs) dwarfed by a huge influx of immigrants in winter (430,000 birds). It is found in freshwater lakes and meadows, as well as estuaries in winter. It breeds mainly in marshes and moorland.

↑ A female (left) and male (right) Teal.

Wigeon

THE GAUDILY COLOURED male Wigeon is one of our very few birds showing pink on its plumage, in this case the breast. It is a very particular duck, which overwhelmingly likes to feed by grazing, usually out of the water, and prefers undisturbed places where this is possible. The short, strong bill is adapted for tugging at grass and other vegetation and cutting it with the bill edges. The Wigeon can also upend or dip its head below the surface while swimming. It is often found in tight flocks, both on land and in flight, and the wonderful wild-sounding whistling of the male is a classic sound of winter birding, especially on the coast.

TOP ID TIPS

A medium-sized duck with a short bill and neck, very pointed wings and a pointed tail. The male's gorgeous plumage looks unreal, with a corn-coloured forehead in addition to the pink breast. The rest of the plumage is pale grey above and below, split by a white line. The female and juvenile are darker and less speckled than other female ducks. The male makes a magical, explosively excited whistle; the female makes a grating rumble.

WHEN AND WHERE TO SEE

Widespread and numerous in the winter, although always localised, preferring larger lakes, reservoirs and gravel pits, as well as floodplains and meadows, saltmarshes and estuaries. Often on nature reserves. Quite rare in summer, breeding in small numbers in lakes and marshes, mainly in Scotland.

↓ A male (front) and female Wigeon.

December's challenge birds

Migrant swans

IT MAY NOT SNOW at Christmas, but you can ensure it's a white one if you look out for swans this month. Everyone knows the Mute Swan, which is common all over the UK, very conspicuous and often tame. But in the cold months of the year, we are visited by two other species: the Whooper Swan, which comes from Iceland (and some from Scandinavia), and the Bewick's Swan, which comes all the way from Siberia. They tend to settle in different places, but sometimes are seen together in mixed flocks. Both species differ from the Mute Swan in that they call in flight but their wings make relatively little noise. In contrast, Mute Swans don't call when flying (hence 'Mute') but make a glorious loud swishing sound with the wings.

TOP ID TIPS

Both species hold their necks straighter than the Mute Swan, which characteristically arches its neck. The bills are more pointed. Closer to, the black and yellow bill colours soon distinguish them from the Mute Swan, which has an orange bill with a black base. Whoopers are bigger than Bewick's, and the yellow on the bill is a pointed wedge shape against the black; in Bewick's the yellow patch is rounder. Both species call in flight, making trumpeting sounds or yelps.

WHEN AND WHERE TO SEE

Whooper Swans are fairly common in the winter in much of Scotland with some large populations further south in England. They arrive in late September and depart in April. Bewick's Swans are much scarcer, being confined to a few floodplains, mainly in England. The first arrive in October, but many don't come until December. They depart in March.

← Bewick's Swan (left) and Whooper Swan (right).

Monthly musings

Where are they now?

You might hear a Robin singing at night at this time of the year, but how do you know that your nocturnal performer isn't a Nightingale? The answer is that all of our Nightingales are currently in West Africa south of the Sahara, in thickets and scrub (where they could, even today, be singing). Meanwhile, Swallows are in South Africa, flying over elephants or giraffes. Cuckoos are in the mighty Congo rainforests, while Swifts could be flying over the East African coast. Some of our Chiffchaffs will be on the coast of Spain, and some of our Goldfinches just a little further north in France. Puffins are in the North Sea or Atlantic, Manx Shearwaters are in the South Atlantic and Kittiwakes are off the coast of North America.

Midwinter singers

It feels like the midwinter doldrums, but believe it or not, several birds sing in December. Robins are audible everywhere, while Dunnocks and Wrens sing loudly. In the morning, you might hear a Song Thrush or Mistle Thrush. Starlings sing on sunny days from rooftops and, amazingly, Skylarks may be heard at the end of the month. More exotically, Cetti's Warblers blast out their loud songs from marshes, and Dippers sing over the sound of rushing streams. And towards the end of the month, the first Great Tits start, signalling the beginning of the rush of song to come.

↓ Cetti's Warblers can be heard singing at this time of year.

Top two must-dos

1 Listen to a flock of wintering geese

Hear a goose on its own, and it just honks. Geese in numbers, however, are lively and entertaining. And geese in multitudes can blow your mind. Seeing and hearing a big flock on a winter's afternoon is one of the finest experiences in British birding.

Canada Geese tend to sully most people's opinion of geese, as they are very common in some areas and can be intimidatingly aggressive when they are nesting, but even they are magnificent in flocks. In many parts of the country, Canada Geese switch between fields where they graze by day to wetlands, such as lakes, where they roost at night. Skeins of them commuting are undeniably eye-catching and this species, which makes very loud honking noises, is a familiar part of the semi-urban winter soundscape.

Many geese travel to Britain and Ireland from the Arctic and sub-Arctic to spend the winter here; we have geese from Russia, Scandinavia, Finland and Iceland, with a few from Greenland, too. They come for the same resources enjoyed by the Canada Geese. They graze or forage

↓ Brent Geese are relatively tame for wild geese and are found in saltmarshes.

→ Wild geese are one of the thrills of winter birding.

on arable fields, grassland or saltmarshes by day, and by night they shift to large waterbodies, including the sea.

The wild wintering geese are localised; the Brent Goose, which is partial to one of the UK's few marine wildflowers, Common Eelgrass, is relatively widespread on estuaries, whereas the rest come to specific areas. As well as Brents, we host Barnacle, Pink-footed, Greylag and White-fronted Geese, along with a tiny population of Tundra Bean Geese.

While geese are fun to watch when feeding, you mustn't miss witnessing the commuting movements, which take place at dawn and dusk. You can find ideal locations on the RSPB website, but for many people, Norfolk is the mecca. If you are in the right place, and have a clear day with a sunset, you might enjoy seeing enormous skeins of Pink-footed Geese plying their way across the sky, at significant altitude. They fly much higher than Canada Geese, the latter only skimming the treetops. Every evening is different, the patterns of lines, chevrons and other configurations written afresh each dusk and dawn.

TOP SPOTS TO WATCH WINTERING GEESE

- The Wash
- Severn Estuary
- Ribble Estuary
- Yare Valley
- Lincolnshire coast
- Loch of Strathbeg
- Ythan Estuary

← Crossbill numbers fluctuate from year to year.

2 Look out for this year's winter nomads

One of the delights of December birding is that, by now, the character of the winter is set. People will be saying things such as 'It's a good year for Bramblings', or better still, 'It's a great winter for Waxwings.'

The avian mix in the non-breeding season is, in many ways, less predictable than in summer. That's partly because, in the autumn, land birds from the continent come here in huge numbers. However, these appearances vary, depending on what is going on in Europe. If Waxwings have enough Rowan berries to eat in Norway and Sweden, they won't come here in any great numbers. The same applies to Long-eared and Short-eared Owls. Their main prey, voles, fluctuate in abundance. If vole numbers crash in Scandinavia, more owls than usual will make it over to the UK. If there are plenty of furry mammals, many owls will stay put, and the numbers of visitors will be modest.

Any bird that depends on unpredictable food sources is prone to these fluctuations in number. When more individuals than usual come here, it is termed an 'irruption'. So, for example, if the crop of mast of Europe's beech trees fails over a wide area, Bramblings will 'irrupt' to the UK, and we see them everywhere. Redpolls depend on birch and Crossbills on spruce. Even Hawfinches will sometimes come over in higher numbers than usual.

One of the side-effects of these irruptions is that the appearance of unexpected birds may cover a wide area. Waxwings might be famous for their love of berries, but they are also famous for turning up in places where birdwatchers don't normally go, such as housing estates and supermarket car parks – the latter are regularly landscaped with berry-bearing shrubs and trees. Waxwings aren't the only ones. Bramblings and Siskins might appear in any garden, and Short-eared Owls may be drawn to any patch of rough ground.

During the winter, food supplies inevitably dwindle locally and birds are forced to seek food in new places. They effectively become nomads, moving in any and every direction, remaining for a while and then moving on. Although they appear in large numbers every winter, Redwings and Fieldfares fall into this category, too. These beautiful birds are highly dependent on berries when they first arrive, and quickly devour supplies. As the weeks pass, they appear in unpredictable waves. They can be almost absent from an area, and then suddenly cloak every hedgerow like colourful Christmas ornaments.

How to help

Shop for nature at Christmas

One thing most of us have to do in December, whether we like it or not, is buy Christmas presents. So, why not put your shopping to work in a way that benefits wildlife, directly or indirectly? One way to help directly is to buy someone a present such as a bird feeder or nest box; the former can be used for the Big Garden Birdwatch in January (see page 19). Indirectly, you can benefit wildlife in myriad ways, not least by buying your cards and gifts from conservation organisations. Or how about buying someone you care about a year of membership to a conservation organisation?

The most straightforward way to help wildlife is simply to give money to the charities committed to wildlife protection. You won't know exactly how it will be spent, but you can rest assured that it certainly helps, however much you can spare. This month, with the year ending, is as good a time as any.

Making your voice heard

What else can we do to help nature during the long, dark evenings of midwinter? It might not seem that we can achieve much, but in fact a little downtime can lead to great things. Some of the most important things we will ever do for wildlife conservation can be done on the sofa, with a phone, tablet or laptop.

→ Buying from the RSPB and other conservation charities can be a Christmas gift to birds.

For instance, have you ever thought of writing to your local political representative? Every communication is taken as representing a significant number of people, so a polite email expressing your support for environmental issues can make waves. This is especially true over pertinent, current local issues. This is a good first step, and going to see your local representative in person is even better. Connecting with local government officials can also be important. Such people can receive a lot of negativity, so understanding and kindness can make a big impression.

Another excellent thing to do is to sign petitions and, if you feel strongly about something, you might consider setting up your own. We all know the power of social media, for better or worse. Why not use it for good? This can often work well if a commercial company gets a crazy notion about removing a birds' roost site because it looks messy, for instance. Companies care about negative publicity.

← Demonstrations can make a big impact on decision-makers.

Myth of the month

The beloved Robin

There can be few other birds or wildlife that cheer us up more than the Robin. The favourite bird of many, the icon of Christmas cards and of good cheer, a gardener's friend and a trusting soul – of course we love them. For centuries Britons have regarded the appearance of a Robin as good luck and, conversely, any harm done to a Robin will bring bad fortune. There is an old saying: 'Good luck to you, good luck to me, Good luck to every Robin that I see.'

This songbird, which performs in the depths of winter, producing carols of its own, is a common species in the vicinity of churches and churchyards. Robins, as we all know, commonly follow gardeners around to snatch up any invertebrates disturbed by digging and weeding. They do the same when graves are dug, leading to the belief that the soul of a departed loved one becomes embodied in the perky bird.

This poem encapsulates the idea.

'It's amazing don't you know?
He visits in the place of a
*   loved one,*
That sadly had to go.
He comes to show they miss you,
Just as you do them,
And they too, long for that day,
That you shall meet again.
They're by your side forever
*   more,*
And that will remain so,
So they send a little robin,
As a way to let you know'

'A VISIT FROM A ROBIN',
AUTHOR UNKNOWN

Photo credits

Front cover tl Richard Bedford/RSPB, tr Edwin Kats/RSPB, bl Craig Churchill/RSPB, br Chris O'Reilly/RSPB; **back cover** t Jules Cox/RSPB, b Andrew Howe/iS; **1** Erni/SS; **3** Kevin Sawford/RSPB; **4** Craig Churchill/RSPB; **5** Chris Gomersall/RSPB; **6** David Tipling/BP; **7** RSPB; **8** Paul Williams/G; **9** Alex Cooper Photography/SS; **10/11/12** Erni/SS; **13** Sandra Standbridge/SS; **14** Alex Cooper Photography/SS; **15** Ben Hall/RSPB; **16** l, tr, br David Tipling/BP; **17** Ed Marshall/RSPB; **18** Paul Sawer/RSPB; **19** Nigel Harris/iS; **20** Erni/SS; **21** David Tipling/BP; **22** Chris Knights/RSPB; **23** Melanie Hobson/SS; **24** Greir/SS; **25** David Tipling/BP; **26** Paul Sawer/RSPB; **27** Giedriius/SS; **28** David Tipling/BP; **29** Nigel Dowsett/SS; **30** Leo Bucher/SS; **31** David O'Brien/iS; **32** Peter Garrity/SS; **33** David Tipling/BP; **34** Ncaan/SS; **35** Kevin Sawford/RSPB; **36** Craig Churchill/RSPB; **37** Jacobi Jayne & Co; **38** Erni/SS; **39** Paul Maguire/SS; **40** Paul Sawer/RSPB; **41** Steve Midgley/SS; **42** Andi Edwards/iS; **43** Ashley Cooper/G; **44** KEMSAB/iS; **45** David Tipling/BP; **46** Alphotographic/iS; **47** Chris O'Reilly/RSPB; **48** Sandra Standbridge/SS; **49/50** Erni/SS; **51** t M Barratt/SS, b Sandra Standbridge/SS; **52** David Tipling/BP; **53** Colin Seddon/SS; **54** Erni/SS; **55** Ian Fox/SS; **56** David Norton/RSPB; **57** Kevin Sawford/RSPB; **58** Malcolm Hunt/RSPB; **59** Digital Wildlife Scotland/SS; **60** David Tipling/BP; **61** Alex Cooper Photography/SS; **62** David Tipling/BP; **63** Erni/SS; **64** t Steve Smith/SS, b Erni/SS; **65** Erni/SS; **66** James Hime/SS; **67** Sandra Standbridge/SS; **68** mtreasure/iS; **69** t David Tipling/BP, bl Ballygally View Images/SS, bc Erni/SS, br Helen J Davies/SS; **70** t Nigel Jarvis/SS, b Kevin Sawford/RSPB; **71** Erni/SS; **72** Steve Round/RSPB; **73** Nick Upton/RSPB; **74** 21csm/iStock; **75** Mark Caunt/SS; **76** Richard Bowler/RSPB; **77** Nigel Blake/RSPB; **78** Alex Cooper Photography/SS; **80** Erni/SS; **81** t Agami Photo Agency/SS; **81** b/**82/83** David Tipling/BO; **84/85/86/87** t Erni/SS; **87** b David Tipling/BP; **88** Mark Caunt SS; **89** Erni/SS; **90** David Tipling/BP; **91** Erni/SS; **92** David Tipling/BP; **93** Ray Kennedy/RSPB; **94/95/96** David Tipling/BP; **97** Sandra Standbridge/SS; **98** David Tipling/BP; **99** Sandra Standbridge/SS; **100/101** t/**100** b/**102/103** Erni/SS; **104** David Tipling/BP; **105** t Nigel Jarvis/SS, b Andi Edwards/iS; **106** Peter Cairns/RSPB; **108** David Tipling/BP; **109/110** Nick Upton/RSPB; **111** Kevin Sawford/RSPB; **112** Genevieve Leaper/RSPB; **113/114** Erni/SS; **115/116** David Tipling/BP; **117** t Alex Cooper Photography/SS, b Erni/SS; **118** David Osborn/SS; **119** Sandra Standbridge/SS; **120** Erni/SS; **121** David Tipling/BP; **122** Erni/SS; **123** t Guy Rogers/RSPB, b Richard Packwood/RSPB; **124/125/126** t /**126** b David Tipling/BP; **127** Sam Turley/RSPB; **128** Nick Upton/RSPB; **129** VDB Photos/SS; **130** Paul Sawer/RSPB; **131** Richard Bowler/RSPB; **132** Erni/SS; **133** Paul Stearman/iS; **134** Mark Caunt/SS; **135** Erni/SS; **136/137** l David Tipling/BP; **137** r Erni SS; **138** Sandra Standbridge/SS; **139/140** David Tipling/BP; **141** t Andi Edwards/iS, **141** b/**142** David Tipling/BP; **143** Erni/SS; **144** Nick Upton/RSPB; **145** Neil Bowman/RSPB; **146** David Tipling/BP; **147** CreativeMoments/iS; **148** Paul Sawer/RSPB; **149** David Tipling/BP; **150** Erni/SS; **151/152/153/154** Sandra Standbridge/SS; **155** David Tipling/BP; **156** t/**157** b Erni/SS, **156** b/**157** a /**158** David Tipling/BP; **159** Sandra Standbridge/SS; **160/162** David Tipling/BP; **163** Nick Upton/RSPB; **164** Jacobi Jayne & Co; **165** David Tipling/BP; **166** Nigel Blake/RSPB; **167** Radovan Zierik/SS; **168** Nick Upton/RSPB; **169** Espen Helland/SS; **170** David Tipling/BP; **171/172** Binson Calfort/SS; **173** Mark robert paton/SS; **174** Helen J Davies/SS; **175** Sandra Standbridge/SS; **176** Mark Caunt/SS; **177** t David Tipling/RSPB, b Ray Kennedy/RSPB; **178/179/180/181** David Tipling/BP; **182/183** Erni/SS; **184** Richard Brooks/RSPB; **185** David Norton/RSPB; **186** Richard Packwood/RSPB; **187** David Tipling/BP; **188** David Kjaer/RSPB; **189** Erni/SS; **190** l, r Sandra Standbridge/SS; **191** Erni/SS; **192/193** Sandra Standbridge/SS; **194** Erni/SS; **195** t Maciej Olszewski/SS, **195** b /**196/197/198/199** b, David Tipling/BP; **199** a Malcolm Hunt/RSPB; **200** Rob Scott/RSPB; **201** t David Tipling/BP, b Erni/SS; **202** Andrew Parkinson/RSPB; **203** Kevin Sawford/RSPB; **204** t Erni/SS, b Nick Upton/RSPB; **205** Anthony Smith Images/SS; **206** David Tipling/BP; **207** Phil Phoenix/SS; **208** Marcin Rogozinski/SS; **209** Ajit Wick/SS; **210/211** David Tipling/BP; **212** l Erni/SS; **212** r/**213** David Tipling/BP; **214** merlinpf/iS; **215** Ernie Janes/RSPB; **216** Roger Tidman/RSPB; **217** RSPB; **218** Ben Andrew/RSPB; **219** David Tipling/BP

Index